Living with Enza

Living with Enza
The Forgotten Story of Britain and the Great Flu Pandemic of 1918

Mark Honigsbaum

Macmillan
London   New York   Melbourne   Hong Kong

First published 2009 by
MACMILLAN
Houndmills, Basingstoke, Hampshire RG21 6XS and
175 Fifth Avenue, New York, N.Y. 10010
Companies and representatives throughout the world

ISBN-13: 978–0–230–21774–4
ISBN-10: 0–230–21774–5

This book is printed on paper suitable for recycling and made from fully managed
and sustained forest sources.

A catalogue record for this book is available from the British Library.

A catalog record for this book is available from the Library of Congress.

10 9 8 7 6 5 4 3 2 1
18 17 16 15 14 13 12 11 10 09

Printed and bound in China

For my father, Frank

contents

list of figures

acknowledgements

I have been fascinated with the 'Spanish' influenza for many years but until I stumbled on Richard Collier's extraordinary collection of letters at the Imperial War Museum I could not see a way of doing justice to the British experience of the pandemic. My first thanks then go to Richard Collier for having the means and wherewithal to collect the testimonies of British survivors before they were lost to history, and to the trustees of the Imperial War Museum for allowing me to quote from the letters freely.

My second stroke of good fortune was in discovering Will Francis at the literary agency Greene and Heaton. It was Will who encouraged me to see the testimonies in the Collier Collection not merely as first-hand accounts of a catastrophic disease event but as a window onto a particular period of British history. In particular, Will worked with me on early drafts of chapter four, helping to contextualize the material and to bring out the conditions on the Home Front – conditions which, I believe, go some way to explaining the stoicism of British survivors and why the experience of the pandemic was so quickly 'forgotten.'

Just as it was important to put the pandemic into a proper social context so it was essential to give readers some sort of scientific envelope with which to make sense of the theories about the origins of the 1918 virus and its unusual pathology. In particular, I am indebted to Professor Peter Openshaw at Imperial College, London, for taking the time to explain to me the various ways in which viruses and bacteria trigger infections of the respiratory tract and for helping me to understand the practical constraints on bacteriologists in 1918.

This book could also not have been written without the help of Professor John Oxford at Queen Mary's Medical School, London. Oxford has been fascinated with the 1918 pandemic for far longer than I and in his pursuit of viral fragments from 1918 has built up an invaluable archive of official reports, letters and other historical documents, many of which he kindly made available to me in the course of my research. He and his colleague, the military historian Douglas Gill, also shared with me the results of their investigations at Etaples – research which has done much to challenge the theory of a Kansas origin and to restore interest in the British and European history of the pandemic.

In addition, I also enjoyed many fruitful conversations with Jeremy Farrar, the director of the Oxford University Clinical Research Institute at the Hospital for Tropical Diseases in Ho Chi Minh City. Farrar's insights into the epidemiology of bird flu provided me with a useful yardstick against which to measure the debate over the origins of the 1918 virus and the supposed 'inevitably' of the emergence of a similarly lethal pathogen in the near future. All of the above also read key sections of the manuscript, drawing my attention to errors and correcting gaps in my understanding of the virology of influenza and the functioning of the human immune system.

The science was not the only area filled with potential minefields. The voluminous military and social history of WW1 was similarly daunting, as was the regulatory context within which key public health officials such as Arthur Newsholme and George Newman operated. For his insightful comments into the social, economic and political conditions on the Home Front in 1918 I am indebted to Nicky Howard; for their comments on the key medical and health officials who feature in my narrative my thanks to Anne Hardy and Jane Seymour at the Wellcome Trust Centre for the History of Medicine at UCL.

Given the complexity of the science and the challenge of melding so many different historical narratives the manuscript inevitably went through several drafts. For helping me to make sense of what should go where – and what I should leave out – my thanks to my mother Naomi Honigsbaum, Susie Parsons, Tam Fairlie, Hillary Anderson and Tom Quick. I should also like to thank my editor at Macmillan, Alexandra Dawe, whose idea it was to include an up-to-date chapter on bird flu and for her thoroughly professional attitude throughout.

Last but not least, I should like to thank my wife Jeanette for her unfailing support and her acute eye for detail. Any errors that remain are entirely my own.

Mark Honigsbaum
June 2008

prologue

Every Remembrance Sunday at precisely 11am – the so-called eleventh hour of the eleventh day of the eleventh month – all activity ceases in Whitehall and a 'perfect stillness,' to recall George V's phrase, envelopes the Cenotaph. In the two minutes of hushed silence that follow, the Queen, the Prime Minister and the crowds of wellwishers gathered in nearby Parliament Square bow their heads as one and give remembrance for the millions of servicemen who died in the killing fields of Flanders and northern France in 1914–18.

It is a rare moment of peace and tranquillity amid the hurly burly of modern life but though Remembrance Day continues to capture the imagination of new generations – the more so, it seems, as the 'Few' have gotten fewer – there is one group of casualties that is rarely if ever mentioned.

The members of this group never wore khaki, flew the King's colours, or squatted in a trench with shells and bullets whistling overhead but in the weeks leading up to the Armistice they were so numerous that their names dominated the obituary columns in *The Times* relegating the names of the 'fallen' to an afterthought. The cause of their deaths was usually given as 'influenza' or 'pneumonia' but the familiar medical terminology does not begin to convey the horror of their experiences. Cruelly for a country that had sent the finest and fittest representatives of its generation to war, the majority were young adults in the prime of life. They mostly died in their own beds without medical care, their lips and cheeks bearing the telltale purple-black marks of heliotrope cyanosis, the condition that marked their final battle with pandemic influenza. For reasons that are still not fully understood, the virus that triggered the 'Spanish' influenza (so-called because Spain, not being a party to the war, was one of the few countries to openly report the spreading plague) provoked a massive immune response that caused victims' lungs to fill with fluids and damaged cells. In layman's terms, people choked to death.

In all, some 228,000 Britons expired in the Great Flu pandemic – nearly as many as died taking Passchendaele during the Third Battle of Ypres in 1917. Worldwide, according to the latest calculations, the 1918 virus killed at least 50 million – more than ten times as many as died in the Great War.[1] The result was that 1918 would be the first year since records began that Britain's death rate exceeded its birth rate. But though you will find monuments to WW1 scattered throughout the British Isles, there is no official monument to this, the greatest disease holocaust that Britain or indeed the world, has ever witnessed. Nor will you find many references to the 'Spanish' influenza in the biographies of politicians and other prominent people who lived through the pandemic, or in post-war poetry or fiction. Indeed, by the summer of 1919, when an inaugural victory parade was held in Whitehall in front of Edwin Lutyens' temporary wood and plaster tomb (the permanent version of the Cenotaph carved from Portland Stone was not unveiled until the following year), the Great Flu was for all intents and purposes already 'forgotten.'

Why this should be so puzzled commentators at the time and has continued to puzzle historians ever since. 'Never since the Black Death has such a plague swept over the face of the world,' commented *The Times* in December 1918 '[and] never, perhaps, has a plague been more stoically accepted.'[2]

My interest in the 1918 pandemic was born in 2005 when I was despatched to Vietnam on a reporting assignment to write about bird flu. A particularly nasty virus known as H5N1 (you may have heard of it) was infecting ducks and chickens in the north of the country and a young man and woman were lying mortally ill in a quarantine ward in Hanoi. I am glad to say that thanks to the wonders of antibiotics and modern antiviral medications – neither of which were available in 1918 – the couple survived, but their illness and the panicked reporting that followed it about the imminence of a new pandemic intrigued me. How, I wondered, had people responded in 1918 when they had faced a real as opposed to a merely hypothetical threat from pandemic influenza? Had their upper lips really stiffened, as *The Times* suggests, or had they besieged doctors' surgeries and town halls demanding assistance? And how did people cope with the pain and grief of burying cherished family members and what became of that grief afterwards?

I decided to see if I could speak to survivors but by 2005 people who had survived the pandemic were not easy to find. Eventually, I was given the name of a wonderful lady called Ada Darwin who had been seven when the 'Spanish Lady,' as she put it, had called at her home in Moss Side in Manchester sickening both her and several of her siblings. Despite her age – Darwin was then aged 93 and living in Chester – her memory was as sharp as a razor. Tragically, both Darwin's baby brother and both her parents died in the pandemic but nearly 90 years on she still had perfect recall, telling me she could 'see' the triple funeral cortège passing by a street near her school 'like a film' in her head. Other survivors interviewed by journalists recalled a skipping rhyme sung by schoolchildren at the time.

I had a little bird
Its name was Enza

I opened the window
And in-flu-enza.[3]

Naively perhaps I expected to find more such accounts in the library but when I searched the extensive bibliography of the 1918 pandemic, though I discovered several excellent global histories – and even a New Zealand history – I could find very little that was specific to Britain or the experience of British victims and their families.

There were only two books that came close. The first was a recent work by Naill Johnson, a Cambridge University academic, describing the demography and geography of the pandemic and its social and cultural impact on Britain.[4] The second was a popular history published by the British journalist-turned-writer Richard Collier in the early 1970s entitled *The Plague of the Spanish Lady*.[5] Of the two, Collier's book looked the most promising – the appendix contained a list of more than 1,000 survivors from all over the world. But as I began turning the pages my frustration mounted. Although Collier's book contained fascinating vignettes describing the impact of the flu in Britain, and in places he had quoted the recollections of British survivors, in the end he had opted for an episodic, global narrative that flitted between London, Madrid, Rome and other international cities. The result was that many of the more interesting letters (Collier had placed advertisements in newspapers all over the UK as well as in several other countries) never made it into his book.

Today, this collection of letters – stored in 15 box files at the Imperial War Museum – represents the most extensive record of ordinary people's experiences of the Great Flu anywhere in the world. For the social historian, the box containing the British correspondence alone is invaluable and I have drawn on it extensively, particularly in chapter four where I describe the impact of the autumn wave on British cities and towns.

These letters amount to what the distinguished medical historian Roy Porter in a famous rallying call on behalf of the medically disenfranchised called a 'patients' view' of history or 'history from below.'[6] But while I like to believe that Porter, were he still alive, would have shared my fascination with the testimonies contained in the Collier Collection, I think that in the end he would have had to acknowledge that a patients' history of the pandemic only gets you so far.

One reason why patients' experiences of the Great Flu were so easy to 'forget' was that historians and social institutions were not interested in recording them. As Johnson puts it, flu was an 'unregarded' and 'overshadowed' killer in 1918, 'a bit player, in the larger story of the Great War.'[7] My aim in this book has been to return the flu to centre stage. To do so I have had to draw on a variety of sources. These include not only survivors' testimonies but the official reports and journals of medical health officers and hospital doctors, as well as investigations into the pathology and 'bacteriology' of influenza published in the *Lancet* and the *British Medical Journal*. In addition I have made liberal use of articles published in both national and local newspapers – many of which resisted the patriotic urge to downplay the spreading morbidity and mortality. Finally, I have examined the records of meetings attended by senior government health officials whose job it was to plan the civilian response, and the diaries, letters and memoirs of politicians, writers, and poets who lived through the pandemic and now and again thought to record their experiences for posterity. What emerges is not simply a social history of the flu but an account of the efforts by medical and scientific researchers to understand the pathology and aetiology of this sphinx-like disease and develop prophylactics – a scientific effort which continues to this day.

In the 1930s Major Greenwood, who had served in the Royal Army Medical Corps during the war (Major was his first name,

not his rank) and later became one of Britain's foremost epidemiologists, remarked that that 'there is some psychological interest
in the fact... that actually the emotional impression created [by
the influenza pandemic] was fainter than that produced by much
less grave epidemiological happenings.'[8]

The central premise of this book is that the story of Britain
and the Great Flu is not merely a psychological curiosity but a
gaping hole in the historical record. It demands to be filled not
only for the sake of the millions of anonymous individuals who
bravely bore its depredations and have since been excised
from history, but because of the very real possibility that
pandemic flu might be about to revisit these isles either in the
form of H5N1 or a related avian virus. According to Britain's
chief medical officer, Sir Liam Donaldson, a new pandemic is
a 'biological inevitability,' hence the government's frantic
stockpiling of antiviral drugs and the drafting of 'pandemic
preparedness plans' envisaging bans on public gatherings and
the conversion of schools and church halls into emergency
infirmaries.

The result is that after nearly a century of burying memories
of 1918, people are once again asking questions about that
pandemic year. Above all, they want to know how doctors,
nurses and ordinary people coped in a world without antibiotics, vaccines and health insurance, and how Britain's
beleaguered National Health Service and the British people
might respond today.

PART I

'No one has yet succeeded in achieving a satisfactory definition of what is meant by influenza, much less a regulation of the usage of the word.'

– F.G. Crookshank, *Influenza: essays*[9]

1

influenza: a primer

On 24 June 1918 the young war poet Wilfred Owen crawled into his Army-issue bell tent in a windswept field near Scarborough and began composing a letter to his mother. Then a 20-year-old lieutenant in the Second Manchesters, Owen had just been deemed fit for duty after a lengthy convalescence in Scotland following an attack of neurasthenia, a nervous condition brought on by the stresses and strain of the war. But as Owen waited in North Yorkshire for the orders that would return him to France and the Front his thoughts were seemingly on another disease entirely.

'*STAND BACK FROM THE PAGE!* and disinfect yourself,' he begins his letter to Susan Owen. 'Quite 1/3 of the Batt and about 30 officers are smitten with the Spanish Flu. The hospital overflowed on Friday, then the Gymnasium was filled, and now all the place seems carpeted with huddled blanketed forms... The boys are dropping on parade like flies in number...'[10]

At first glance, Owen's bold capitals and self-conscious italics read like genuine alarm. But as the next passage makes clear Owen is being ironic and far from taking the disinfectant measures seriously, considers influenza something of a joke. 'The thing is much too *common* for me to take part in. I have quite decided not to! Scottie [a regimental friend], whom I still see sometimes, went under today, & my servant yesterday. Imagine the work that falls on unaffected officers.'[11]

Owen's wry remarks, though calculated to amuse, were typical of British attitudes to influenza that summer. Though at times influenza can resemble a plague, it is most usually regarded as benign and unthreatening. This misconception about influenza results from our tendency to confuse it with the common cold. But while the symptoms of a cold can include a sore throat, cough, and headache, as well as muscle aches and fatigue, colds are caused by rhinoviruses or coronaviruses and only rarely result in serious respiratory complications. Influenza on the other hand is caused by viruses in the Orthomyxoviridae family. Some of these viruses have specific adaptations which enable them to penetrate deep into the respiratory tract and the lungs, and even a mild infection can cause intense muscle and joint pains, headache, and prostration – hence the French term *gripe* from the verb *gripper* meaning to 'grasp or hook.' In the case of pandemic strains of influenza, they can also trigger grave pneumonic complications in otherwise healthy young adults. Having said that, Owen's lack of concern is understandable as the majority of influenza victims usually recover within a week to ten days. Even the less familiar pandemic variety is not an automatic death sentence – up to half a population may be affected but in the developed world at least the case fatality rate rarely exceeds 3 per cent. By comparison a disease like cholera, if untreated, typically has a case fatality rate of 50 per cent and is truly fearful to observe, producing agonizing retching and diarrhoea that leaves its victims dehydrated and dying in watery rice stools.

It is the apparently benign nature of flu, coupled with its familiarity and ubiquity, that breeds complacency. But the ubiquity of flu is exactly why we *should* fear it. Even where outbreaks do not reach epidemic proportions, influenza infects so many people that even the mildest viruses almost always kill. In Britain, for instance, the Department of Health estimates that normal 'seasonal' influenza, which occurs predominantly during a six to eight week period in winter and affects between 5–10 per cent of the population accounts for 12,000 deaths annually, the majority of its victims being the very young and elderly or those with underlying chest and heart conditions. But during the 1918–19 pandemic, according to some estimates, up to a third the British population, or some ten million people, were affected.[12] Although the case fatality was just 2.5 per cent, the result was 228,000 deaths – 25 times more than in a normal flu season.[13]

All mammalian influenzas begin as viruses of birds but though we know some of the ways in which these avian viruses mutate so as to infect pigs and humans we still do not know the precise evolutionary and environmental pressures driving the mutations. Nor do we know whether the 1918 virus originated in a bird or pig, or some other as yet unidentified animal host, and precisely how it became infectious in people (for further discussion of these issues see chapter six).

Without intact viral genetic material definitive retrospective diagnoses of past influenza epidemics are impossible, but from historical records it's safe to say that highly contagious, acute respiratory illnesses have afflicted mankind since the beginning of civilization. Both the Roman historian Livy and Hippocrates, the Greek father of medicine, describe a disease that sounds very much like influenza sweeping northern Greece in 412 BC and in 1411 the French applied the name 'tac' to an epidemic disease whose features included a dry hacking cough accompanied by a chill and extreme lethargy.

In September 1485 a similarly mysterious disease suddenly appeared in London. Dubbed the *sudor Anglicus*, or the 'English sweat,' it was thought to have been imported to the English capital by Henry Tudor following his victory over Richard III at

Bosworth Field. According to contemporary medical chron-
iclers, the symptoms included 'grete swetyng,' a 'contynual
thurst' and an intense headache or 'pricking [of] the brains.'
But the most unusual characteristic was the suddenness of the
attacks and the way that the disease seemed to target the rich
and better-off members of society. By the time the outbreak
had subsided in late October, it had dispatched two London
mayors and four city aldermen.[14] Writing in 1551 after the fifth
and final visitation of the sweats to the British Isles (the other
outbreaks occurred in 1508, 1517 and 1528), John Caius,
physician to Henry VIII and Edward VI, said that those who
were 'sore with peril of death were either men of wealth, ease
or welfare, or of the poorer sort, such as were idle persons,
good ale drinkes and taverne haunters.' He also had little
doubt that the disease was of foreign origin, blaming it
on Henry's army of French mercenary archers recruited in
Rouen.[15]

However, the English Sweat does not appear to have been
accompanied by a cough or other marked respiratory symptoms, a key hallmark of pandemic influenza. The same cannot
be said of '*coqueluche*' (from the French term for a monk's
hood because those so afflicted wrapped themselves in a hood
to contain their shivering) a severe pulmonary disease which
in 1510 swept across Europe from Africa 'not missing a
family and scarce a person'. The disease, which was accompanied by earthquakes and a volcanic eruption in Iceland, was
almost certainly influenza, the key symptoms being a sharp
headache, chill and 'a terrible tarring cough' so violent that
one observer reported many patients were 'in danger of
suffocation.'

In 1562 another disease dubbed the 'Newe Acquayntance'
swept through Queen Mary's court in Edinburgh at the end of
November – the classic period for the onset of seasonal winter

flu. Lord Randolph writing to Lord Cecil from Edinburgh described it as:

> ... a plague in their heads that have yt, and a soreness in their stomaches, with a great cough, that remayneth with some longer, with others shorter tyme... the Queen kept her bed six days. There was no appearance of danger, nor manie that die of the disease, except some olde folks.[16]

It is difficult to say when the first true influenza pandemic occurred. Many medical historians argue the most likely candidate is the disease that swept northward from the Mediterranean to the Baltic in 1580. Within six weeks it had infected every country in Europe and was so ubiquitous that contemporary commentators remarked that 'hardly the twentieth person was free of the disease.' In Rome alone it left 9,000 dead and during the war against the Desmonds in Ireland it decimated the ranks of the English Army. 'For three daies,' reported one observer, 'they laie as dead as stockes, looking still when they should die; but yet such was the will of God that ... they all recovered.' There are also contemporary accounts from Asia and Africa, suggesting that this was a disease that affected people on every continent at the same time, which is the very definition of a pandemic (by contrast, an epidemic is merely a disease that affects people at the same time in a particular community, institution or country).

In the 17th century we can pinpoint at least three further candidates, including an epidemic that struck England, Ireland and Virginia in 1688, the year of the Glorious Revolution, causing people to die 'as in a plague.' Five years later, in 1693, influenza was again raging in Europe, attacking people of all conditions and ages, including 'those that were robust and hardy, as well as those that were weak and tender.'[17]

At this time it was common to ascribe outbreaks of epidemic catarrhs and coughs accompanied by chills and great lassitude to volcanic eruptions and the passage of meteors through the night sky – a conception that persists in some scientific circles to this day.[18] Indeed, the term influenza most likely derives from the Italian phrase *influenza coeli*, meaning 'influence of the heavens.' However, by the mid-18th century it had become more common in Italy to talk of *influenza di freddo* – 'influence of the cold' – and it was probably in this sense that the term first entered the English language in 1743.[19]

In the 18th century it is thought that there were at least three, and possibly as many as eight, influenza pandemics and once again contemporary commentators noted the democratic nature of the victims. During the 1732–33 outbreak in Plymouth, for instance, it was reported that 'scarce anyone had escaped,' while in 1781–82 it is said that two-thirds of Rome's population were attacked.[20] However, these outbreaks pale in comparison with the 1836–37 and 1847–48 pandemics. The first, which swept west across Europe from Russia, claimed 3,000 lives in Dublin alone – more than the 1832 cholera epidemic – and in London the 1847–48 pandemic triggered 5,000 more deaths than in a normal flu season. But while a London doctor described the 1837 pandemic as one of the 'most direful scourges' he'd ever experienced, and in 1847 the ravages of influenza were once again compared to cholera, these early 19th century outbreaks apparently occasioned little overt panic.

'Do you know what it is to succumb under an insurmountable daymare ... an indisposition to do anything, or to be anything; a total deadness and distaste; a suspension of vitality,' asked the English essayist Charles Lamb laconically in a letter to his friend the Quaker poet Bernard Barton in 1824. 'This has been for many weeks my lot, and my excuse; my fingers drag heavily over this paper, and to my thinking, it is three and twenty furlongs from here to the end of this demisheet.'[21]

Thomas Carlyle, writing to his sister at the height of the 1837 pandemic, was similarly non-plussed describing influenza as 'a dirty, feverish kind of cold.' In common with his contemporaries Carlyle blamed the outbreak on excess humidity, in particular the damp fogs that arrived in London from the northeast. 'Printing offices, Manufactories, Tailor shops and such like are struck silent, every second man lying sniftering in his respective place of abode,' he complained.[22]

In fact, Carlyle's observation is based on a misconception. Unlike the viruses which cause colds and which spread primarily by direct person-to-person contact, influenza viruses spread aerially, usually in small droplets expelled when someone coughs or sneezes, and tend to be more stable in cool, dry conditions. At room temperature, influenza peaks at a relatively low humidity – around 20 per cent – but at over 80 per cent humidity the droplets become too heavy and fall to the ground. Researchers have also discovered that at around 5° Celsius the virus transmits for about two days longer than at 20°.[23] In other words, the children's rhyme about 'in-flu-enza' entering via an open window may be spot on.

***

By the 1890s many Britons had lulled themselves into thinking that the wonders of medical science could vanquish any foe, no matter how microscopic. In Germany Robert Koch had isolated the bacilli of tuberculosis and cholera while in Paris Louis Pasteur had developed vaccines against anthrax and rabies and discovered the process of pasteurization. But, for all their achievements, neither had the least notion what a virus was.

True, bacteriologists had begun to suspect that some diseases might be spread by what they called 'filter-passers' – microbes small enough to pass through the filters used in the laboratory isolation of bacteria – but as far as most scientists were concerned there was no mystery about influenza. Influenza, they thought, was spread by a bacillus like cholera, a belief that seemed to be confirmed in 1892 when one of Koch's most brilliant disciples, Richard Pfeiffer, announced in Berlin that he'd isolated the infective agent, *Bacillus influenzae*, later known as Pfeiffer's bacillus. In fact, *Haemophilus influenzae,* as Pfeiffer's bacillus is known today, is a fellow traveller and like other bacteria commonly found in the throats and lungs of influenza patients (principally pneumococci, streptococci and staphylococci) is not the primary cause of the disease.[24]

It would not be until long after WW1, in 1933, that a team of British scientists working for the Medical Research Council (MRC) would demonstrate that influenza was actually transmitted by a virus, and it was not until the 1940s that researchers would see the influenza virion for the first time. Viewed through an electron microscope it resembled nothing so much as the surface of a dandelion bristling with tiny spikes and mushroom-like spines. The spikes, we now know, are made of a protein called haemagglutinin (HA) which derives its name from its ability to agglutinate to red blood cells. When a person inhales an air droplet containing the virus it is these spikes that stick to the receptors on the surface of the cells in the respiratory tract, much as a prickly seed case catches on the fibres of clothing in tall grass. The square-headed mushroom-like protrusions, fewer in number, are a powerful enzyme, neuraminidase (NA). It is the combination of these proteins and enzymes that enable the virus to evade the body's immune defences, much as a burglar equipped with a set of skeleton keys prises open a lock.

There are three types of influenza viruses: A, B and C. Type C viruses rarely cause disease in humans, while type Bs produce classic winter flu. Neither pose an epidemic or pandemic threat. That is the preserve of type A viruses, the primary reservoir of which is wildfowl such as geese and ducks. Unlike the building blocks of the human body – the double-stranded helix spirals of DNA (deoxyribonucleic acid) – influenza viruses, including type A viruses, are composed of eight delicate strands of RNA (ribonucleic acid). It is these RNA strands that code for the proteins and enzymes on the surface of the virus and determine the particular configurations of H's and N's.

Unfortunately, RNA does not possess an accurate proof reading mechanism. During replication, when the virus invades and colonizes animal cells, manufacturing hundreds of thousands of copies of itself, the RNA makes small copying errors, resulting in genetic mutations to the surface antigens – the combination of H's and N's which dictate the production of antibodies and the body's immune response. These mutations are known as 'antigenic drift'.

In addition, type A viruses can also 'swap' or reassort genetic material with other viruses. This process, known as 'antigenic shift', usually occurs when an avian or swine strain of influenza A exchanges genes with a human version of the virus, producing a completely new subtype. Inside the host, the eight RNA gene segments are shuffled randomly, like the symbols in the window of a slot machine. The result is a new virus that codes for proteins that may be new to the immune system and to which the body has no antibodies. It is these strains that historically have been the cause of pandemics.

The genetic identity of pre-modern pandemic strains of influenza are lost to history, but in the modern era there have been four major shifts. The first, H1N1, is the name given to the 1918 strain and is currently found only in pigs. This was also the dominant strain in humans until 1957 when a new subtype, H2N2, suddenly emerged in Asia triggering a pandemic that killed an estimated one million people worldwide. In 1968 there was a third shift, with the sudden emergence of H3N2 in Hong Kong, which also caused some one million global deaths. This is the strain currently circulating in human populations today. In addition, in 1977 there was a puzzling re-emergence of H1N1 – probably as the result of laboratory error – but it failed to trigger a pandemic or replace the prevailing H3N2 subtype.

Since the middle of the 19th century most pandemics have been thought to originate in the 'silent spaces' of Asia and the Far East – a suitably broad and imprecise sweep of territory stretching from southern China and southwestern Siberia to Kazakhstan and the Urals. But while many flu's have been described as 'Asiatic' they have also been attributed to other peoples and nations, including in 1918 the Spanish, and in 1889 the Russians.

The 'Russian' flu pandemic came in three waves over a period of as many years sweeping west from Bokhara in Tsarist-controlled Uzbekistan to St Petersburg via the new Trans-Caspian railway. From St Petersburg commercial and diplomatic travellers conveyed the disease to the Baltic ports while rail passengers spread it to Warsaw, Berlin and Vienna. By December the disease was in Paris, and by the early winter of 1890 it had reached London from where it was disseminated to the rest of Britain. The speed with which infections spread and the way the disease appeared to follow the lines of communication and commerce attracted widespread comment, as did the democratic nature of the attacks. 'From emperors to potboys, no one has been exempt,' commented one journal, 'we have all ached in common.'[25]

Figure 1.1 'Everyone has influenza.' Wood engraving for **Le Grelot**,
12 January 1890.
Credit: Wellcome Library, London

In London alone it is estimated that the first wave of Russian flu sickened 10–15 per cent of the population or in excess of 400,000 people. The official death toll was low – just 4,523 recorded deaths from influenza for the whole of England and Wales. But medical health officers, whose job it was to monitor local disease outbreaks and report back to government, noticed that this was not the whole story and that when deaths from associated respiratory complaints, such as bronchitis and pneumonia, were taken into account, the true excess death toll was 27,000. Leading London teaching hospitals such as St Bartholomew's in the City were particularly hard hit with doctors arriving at morning surgery to find upwards of a 1,000 patients already awaiting treatment. This flu was nothing like a cold but seemed to strike at the heart of late Victorian industrial society, seizing men, the main family breadwinners, in their places of work. Some patients could even name the exact hour of an attack saying, 'I went to work all right this morning, but was suddenly taken ill at eleven o'clock, and had to leave off.' The disease was characterized by extreme prostration, weakness and nervous depression. Frontal headache, pain in the eyeballs and muscular pains were also common, and, in the worst cases, a peculiar 'creeping pneumonia.'[26]

Public apprehension deepened when it was reported that two soldiers had died at the Guards Hospital in Rochester Row and that Lord Salisbury, the Prime Minister, was laid up at his country seat at Hatfield. The Queen asked for regular updates and saw her Prime Minister recover. But nearly a year later, in the spring of 1891, the flu returned sweeping south along the Pennines to Sheffield from where it was carried to London and back to the Houses of Parliament. One of the first victims was the Archbishop of York who had been in Westminster pressing the case for child insurance. As Lords, MPs and even the Prince of Wales were taken ill, fumigation squads were brought in to spray the Houses of Parliament in the hope of chasing the 'microbe' from Westminster's insanitary nooks and crannies. In Sheffield, meanwhile, the death rate in the first week of May reached 70 per 1,000 – the highest rate in the town's history, exceeding even the 1832 cholera epidemic.[27]

By the end of the Whitsuntide holiday, however, the epidemic was on the wane, and although the Russian flu returned for a third time in the autumn and winter of 1891–92, this time it was nothing like as severe. Nevertheless, the spring 1891 wave alone had caused nearly 58,000 excess deaths, and including the follow-on waves in 1893 and 1894 it is estimated that more than 100,000 Britons lost their lives.[28]

The Russian flu pandemic awakened interest in the disease as never before. Whereas before 1889 most doctors regarded flu as relatively mild and harmless, after 1889 medical professionals had little choice but to take it seriously. In particular, doctors were struck by the association between the Russian flu and what they described as 'a low and insidious form of pneumonia.' It was these pneumonias, which lodged deep in the lobes of the lungs so as to mimic the symptoms of catarrh, that had caused most of the deaths, though whether these 'lobar pneumonias' were due to the 'microbe' itself or secondary bacterial infections no one could be certain. The other alarming feature was that while, as in a normal seasonal flu, the elderly and the very young had been the worst affected – a graph of mortality by age showing the usual U-shaped curve – men and women aged 30–40 had suffered an abnormally high number of lung complications. Not only that but after the initial attack had passed and patients were thought to be on the road to recovery 'nervous complications' had supervened sparking depression, lethargy, and, in some cases, suicides. Finally students of epidemiology observed how the pandemic had taken the form of three distinct waves of infection with the second wave – which arrived in the spring of 1891 – being more severe than the first, and the third and final wave in the winter of 1891–2 being the least severe.[29]

In 1918 this pattern of a mild primary wave followed by a severe secondary wave would be repeated but this time the mortality would fall disproportionately on young adults – men and women aged 20–40 – producing a W-shaped mortality curve with the tallest peak in middle life. This was a phenomenon that had never been seen before. Nor has it been repeated since. But the most alarming feature was the manner in which people died. 'Spanish' influenza struck suddenly and without warning: one moment a person was up and about, the next they would be lying incapacitated coughing up greenish-yellow sputum. In some cases, a frothy, blood-stained fluid gushed from their nose and mouth. As pneumonia set in their temperature would soar to 40 or 41° C and they would slip into a delirium. The final stage came when their lungs filled with fluid prompting their heart to leech oxygen from the blood vessels supplying the head and feet. This was the condition known as heliotrope cyanosis. It must have felt like drowning. Typically, a blue or dark-purple stain would spread across the lips and cheekbones. Then the victim would turn a mahogany colour and die. In the words of one historian of the 1918 pandemic, the flu 'turned people the color of wet ashes.'[30]

But in January 1918 such horrors still lay in the future. After nearly four years of 'total warfare' during which Britons had suffered Zeppelin raids, food and fuel shortages, and the loss of hundreds of thousands young men in the 'killing fields' of Flanders and northern France, the prospect of being struck down by an epidemic disease, let alone something as familiar as the flu, was not uppermost in most people's minds. In those cold and hungry years pneumonia and other respiratory diseases, such as tuberculosis, were a far more present threat. Perhaps that is why when reports of a strange new epidemic disease first emerged from Spain in May 1918 the *Times*'s Glasgow correspondent reported that the man in the street dismissed the threat 'and cheerfully anticipated its arrival here.'[31]

In the United States, however, attitudes were different. There, General Pershing was on the point of dispatching a million 'doughboys' to Europe and US Army medics emboldled by their victories over yellow fever in New Orleans were on the look out for unusual disease outbreaks. The result was that when hundreds of soldiers at an Army encampment in Kansas were suddenly struck down by a flu-like disease accompanied by severe pneumonia in March 1918 a telegraph was sent to Washington. It was the first report anywhere in the world of what would later erroneously be labelled 'Spanish' influenza. Within weeks the flu had spread to other US Army camps and by April it had reached Bordeaux, courtesy of the American Expeditionary Force landings at Brest. By May 'three-day fever' and 'grippe' was rife throughout the Allied lines and the Germans were complaining of 'Blitzkattarh.' By now there were also reports from Scapa Flow in Scotland that 9,000 seaman – 10 per cent of the total fighting force of the British Grand Fleet – were laid up. Then, in June, came rumours that some 3,000 Tommies had been taken by hospital trains to a massive British Army hospital complex on the northern French coast near Boulogne. The official diagnosis was 'P.U.O.' – Pryexia of Unknown Origin – but most medics at the camp were convinced it was the same disease that had accompanied the AEF to Bordeaux. What is more they thought they had seen a very similar disease two years before.

2
prelude: Etaples, winter 1916–1918

In the winter of 1915, as Germany was reinforcing its extensive trench system at Ypres, the British Army began the construction of what would be the largest hospital camp any government has ever built abroad. The site selected for this ambitious project was Etaples, a small village 15 miles south of Boulogne, labelled a 'sand heap' by a leading British medical journal because of its proximity to the windswept dunes of the northeastern French coast.

In fact, Etaples had much to recommend it. First, the town was surrounded by miles of open paddocks perfect for the erection of hospital tents and huts. Second, it bordered the railway line running southwards to Abbeville and the Somme, meaning that soldiers, once recovered from their wounds, could be quickly returned to the Front with the minimum of fuss. Third, Etaples' proximity to Boulogne meant that reinforcements arriving from Britain would have the shortest of sea crossings – an important factor given the large number of German submarines and mines lurking in the Channel. But the main *raison d'etre* for Etaples and the key factor that drove its growth was the sheer scale of the military confrontation on the Western Front and the diseases that confrontation brought. Never before or since have so many men and so many microbes co-existed in the same place for so long, or with such devastating consequences.

By the winter of 1917 some two million Allied troops were dug in across northern France and Flanders – not just British and French, but Portuguese, Canadians, South Africans, Australians, New Zealanders, West Indians, Punjabis and North Africans. Just behind the lines, helping to ferry munitions, food and forage from the Channel ports to the trenches, were a million or more Chinese recruited by the British in northern China. As in any war, the main medical concern was tetanus and graver forms of sepsis developing in open wounds – a considerable problem in WWI given the use of high-explosive shells, rapid-firing machine guns and the legions of bacteria that festered in the Flanders mud. Then there were the usual worries about diseases associated with close confinement in unsanitary conditions, such as dysentery and typhoid. But in Flanders Army doctors were also presented with diseases they had never encountered before and which are seldom seen in war, such as trench foot and trench fever. The first, the result of the ceaseless exposure to the cold oozing mud, caused men's toes to become gangrenous until they literally rotted and fell off. The second (though this was not discovered until after the war) was a form of relapsing fever transmitted by a louse, *Pediculus humanus*, that was endemic to the trenches.[32] Trench fever was also associated with another indeterminate disease known as 'war nephritis,' a condition associated with headache, fever and kidney damage and which, like trench fever, was new to WWI. In addition, following Germany's decision at Ypres in April 1915 to ignore mutually agreed conventions prohibiting the use of chlorine and other poisonous gases, the Royal Army Medical Corps (RAMC) also had to contend with never-before-seen broncho-pneumonic conditions associated with the gas damage to men's lungs.

However, all these would be overshadowed by influenza, a disease that in the spring of 1918 swept through the lines so suddenly and with such ferocity that it startled even doctors who'd served in Gallipoli and Salonika and witnessed wards overflowing with amoebic dysentery and malaria cases. Indeed, until May and June of 1918 battalion sick rates from infectious disease had been relatively low. The flu changed all that, sickening 226,000 men in May to August 1918 and a further 87,000 between October and the end of the hostilities. It was, concluded British Army medical statisticians writing in 1933, 'a world-wide epidemic which no medical service could control, and which laid low both friend and foe alike.'[33] But although doctors were powerless to prevent Spanish influenza, in theory Etaples should have been well equipped to deal with the onslaught. In fact, as we shall see, not only were the camp's hospital facilities inadequate for the demands that were placed on them, but the conditions at Etaples may have amplified and, according to some scientists, even provoked the critical adaptations in the virus.

***

Today, Etaples is little more than a dot on the map, a round-about en route to the casinos and beaches at Le Touquet and Paris Plage (the only evidence of the formerly extensive British military presence is a huge Lutyens-designed cemetery dedicated to fallen soldiers of the Commonwealth on the outskirts of the village). In 1917, however, Etaples presented a very different sight. Arriving in northern France in the autumn of that year, Vera Brittain, then a young nurse in the Red Cross's Voluntary Aid Detachment (VAD), was struck by the 'huge area of the camps' with the main line from Boulogne to Paris running 'between the hospitals and the distant sea.'[34] Built on a series of low-lying meadows surrounded by 'humped sand-hills' Etaples seemed to her not so much a camp as a small city.

Row upon row of tented sleeping quarters for the doctors and Alwyn huts for the nurses stretched across the fields on either side of the railway line. At the centre were 12 long wooden huts. Decorated with scarlet and yellow nasturtiums, these were the hospital wards with beds capable of accommodating up to 22,000 men. In addition, on rising ground to the east of the railway line, the Army had built three infantry depots – each one a separate street lined with huts and tents and surrounded by barbed wire – as well as training grounds, a firing range, a cemetery, laundry and two post offices. Finally, Etaples boasted extensive stables for the thousands of horses requiring veterinary attention, as well as piggeries for food.

Figure 2.1 The piggery at Etaples constructed by the Army's Directorate of Works in 1917.

Credit: by permission of the Imperial War Museum

Adding to the camp's farmyard ambience were pet ducks, geese and chickens bought by men at local markets.[35]

On any given day a hundred trains laden with soldiers and supplies might pass through the sidings at Etaples. In addition, there was a constant passage of men in wheeled transports. Indeed, one historian of the war calculates that at the peak of its activity in 1917 and 1918, on any given day Etaples would have contained some 100,000 men and women, and that between June 1915 and September 1917 more than a million officers and men passed through the reinforcement camp on their way to the Front.[36]

At first Brittain was enchanted by the vista, remarking that thanks to its proximity to the sea and the mists that swept across the marshes Etaples was lit by a 'vivid golden-yellow light' that turned to a 'dazzling green' in winter. But after a few weeks at the camp her awe turned to dismay as Etaples was inundated with 'perpetual convoys' of men with 'fear-darkened' eyes and even more dishevelled detachments of soldiers returning with fatal wounds – casualties of the fierce fighting around Passchendaele. In all, the British would lose 250,000 troops during the five-month offensive on the village in the Ypres salient before the Canadians finally took it from the Germans in November 1917.

According to some scientists, the immense movement of soldiers through Etaples, coupled with the large numbers of pigs, chickens and horses at the camp provided the perfect conditions for the emergence of pandemic influenza. The other critical factor may have been the presence of mustard and other mutagenic gases. Following the immense loss of life at the Battle of the Somme, the use of poison gas had become a strategic imperative for both the Allies and the Germans. Gases like phosgene and chlorine were not only capable of disabling and killing on contact, they also acted as soil contaminants denying valuable ground to the enemy. In all, it is estimated that some 150,000 tonnes of poison gases – the equivalent in weight to the cargo of a modern super-tanker – were dumped on the killing fields of Flanders and northern France during the last two years of the war, saturating the soil to the point where it became impossible for attacking troops to hold territory without large numbers having to retreat to hospitals with suppurating blisters and damaged lungs and eyes. As one officer recalled, the 'gas was everywhere, in clothing, food and water. It corroded human skin, internal organs, even steel weapons.'[37]

By 1917 gas attacks were part and parcel of trench warfare. Some gases killed instantly, but others had a delayed effect resulting in many men staggering back to camp where they contaminated their colleagues. In the case of Etaples, there is also evidence raw recruits were deliberately exposed to poison gas as part of their training with one Tommy recalling how on arrival at the camp he was instructed to crawl through a chamber into which gas had been released: 'The instructors used to say, "Just to get used to the real thing when you get there."'[38]

The most mutagenic gas of all was mustard gas (dichlorethyl sulphide). John Oxford, Professor of Virology at Queen Mary's Medical School, London, and the military historian Douglas Gill estimate that some 12,000 tons of the gas were dumped on the Western Front following its introduction in 1917 and that in all mustard gas accounted for 400,000 casualties.[39] At this remove in time, the precise impact of mustard and other gases is difficult to quantify. To date, no one has attempted to recreate the conditions at Etaples under laboratory conditions so the hypothesis that – as Oxford and Gill put it – these agents may have prompted 'stepwise mutational changes' in the influenza virus remains untested. But in combination with the bitterly cold conditions that prevailed on the Western Front in the winter of 1917 and the stresses and strains of war, it is possible such contaminants would have lowered mens' resistance.

'There were particular circumstances after the Battle of the Somme – so many young people, so much distress, so much gas – that would lead to the transmission of a virus like influenza,' argues Oxford. '[Certainly] the crowding together of men in a military camp like Etaples would have enhanced the transmission of the virus much more rapidly than in a civilian population.'[40]

Although influenza was not a reportable disease in Britain at the time, from camp medical records we know that influenza was widespread at Etaples in 1916. Furthermore, that winter Army medical officers also observed the emergence of a strange new respiratory infection – an infection that in retrospect the same medics would identify as essentially identical to the 'influenzal pneumonias' they had seen during the so-called Spanish influenza outbreak of 1918–19. The medics labelled the disease 'purulent bronchitis' because it was characterized by masses of purulent blood-streaked and yellowish-green sputum. As in 1918–19, there was also another unusual feature: heliotrope cyanosis.

***

The lungs are the lightest organs in the body. In someone with healthy lung function, the tiny alveoli at the end of the branching network of bronchi and bronchiole keep the tissue as buoyant as a sponge. But with the onset of pneumonia, pneumococci multiply faster than the patient's immune system can produce antibodies to the bacteria, filling the tiny air sacs with blood and other fluids and causing the alveoli to harden into a consolidated mass. The lungs are so elastic and airy that if you were to put them in a bathtub they would sink to the bottom. For the patient this process of consolidation is cata-strophic. As the alveoli shut down, the amount of oxygen exchanged into the bloodstream is drastically reduced and the skin loses its natural hue turning lavender and dark purple. This is the condition that doctors call heliotrope cyanosis. In the case of some patients with pneumonia, the immune system will fight off the bacterial invaders, but in the case of viral pneumonias the flooding of the alveoli may be caused by the immune system itself. Either way, in 1918, without antibiotics to kill the bacteria or antiviral drugs to reduce the viral load, there was little that doctors could do except make patients comfortable and wait. If the cyanosis disappeared and a healthy pink hue returned that meant patients were usually on the way to recovery. If, however, the discolouration darkened that meant the prognosis was bad.

During the war doctors saw many different kinds of respiratory disease. The commonest was bronchitis or catarrh, but during prolonged cold snaps some of these catarrhs might be labelled 'epidemic.' In addition, the war years saw a marked increase in tuberculosis cases after nearly a decade of steady declines, as well as cases of what doctors called 'fulminating pneumonia.' Purulent bronchitis, however, was a new term. It was coined by an RAMC medical officer, Lieutenant J.A.B Hammond, and his colleagues, Army pathologist Captain William Rolland and T.H.G. Shore, the officer in charge of the Etaples mortuary, in a paper that appeared in the *Lancet* in July 1917.[41] The paper detailed the results of a series of bacteriological exams on the sputum and pus of 20 men who'd died or recovered from the disease at Etaples the previous winter. To protect the men's identity, Hammond listed only their rank and the first initial of their surnames. But thanks to sleuthing by Gill we now know that patient eight, listed only as 'Pte. U.' in the *Lancet*, was Harry Underdown, a 21-year-old hay trusser from Ashford in Kent. A private in the 12th Battalion, Queen's (Royal West Surrey) Regiment, Underdown had been called to active service in April 1916 but had almost immediately been hospitalized with tonsillitis. Following his recovery in August 1916 he was sent out to France but within a few weeks he was back in hospital, this time suffering from shell shock. Underdown spent the autumn of 1916 at Bagthorpe Military Hospital in Nottingham, where he was prescribed 'rest and bromides,' before being discharged in November and returned to active duty. He crossed to France and Etaples in February 1917, dying of 'widespread broncho-pneumonia' at No. 24 General Hospital on the 21st of the same month.

Though the *Lancet* paper details just 20 cases of purulent bronchitis, Underdown was actually one of 156 soldiers to die of the disease at Etaples in February and March of 1917. It was Sir John Rose Bradford, the consultant physician at Etaples and a future president of the Royal College of Physicians, who first noticed the unusual symptoms and suggested Hammond make a special study of the outbreak. An enthusiastic advocate of laboratory based research (towards the end of the war Bradford would sponsor a series of failed experiments with monkeys in an attempt to isolate influenza and the viruses of other common infectious diseases), Bradford had arrived at Etaples in 1915. At first he complained that the cases which came under his purview were 'very narrow' and there was little to sustain his 'professional interest.'[42] The outbreak of purulent bronchitis, however, changed that. In acute cases the patients' temperatures had shot up to 103 degrees, before dropping off a few days later. But many patients experienced great difficulty in breathing and some were so panicked by the feeling of breathlessness that they struggled 'wildly' and leapt out of

bed. Initially, the cases struck Bradford and Hammond as ordinary lobar pneumonia, with the sound of cracking rales clearly
audible at the root of the patients' lungs. But the expectoration of masses of purulent pus was an unusual feature,
particularly as the symptom was more usually associated with
bronchitis. In addition, the most acutely affected men turned
cyanotic.

According to the *Lancet* report, this condition, which only
became apparent when the patient ceased to bring up the
purulent expectorations, usually marked the end stages of the
disease and was detectable by a 'duskiness' of skin tone
turning to a distinctive blue discolouration that spread from
the lips across the cheeks and ears. Although the cyanosis
could be temporarily relieved by oxygen therapy, patients
found the oxygen masks uncomfortable and frequently tore
them off. In nearly every case, the Etaples researchers reported,
death from 'lung block' followed soon afterwards.

The most peculiar features of all, however, only became
apparent during autopsy. In a normal case of lobar pneumonia, pathologists would expect to see damage to only one
of the lobes of a patient's lungs but in the Etaples cases there
was also widespread bronchitis. On being sliced open the
smaller bronchi oozed a thick yellowish pus and were also, in
some cases, found to contain *H. influenzae* and other bacteria.
Indeed, of the 156 soldiers who died at Etaples in the winter of
1917, nearly half – 45 per cent – were found to have purulent
excretions blocking the smaller bronchi. These features were
so marked and the symptoms of the disease so distinctive,
argued Hammond and his colleagues, that they constituted a
'definite clinical entity' – hence their decision to name the
disease purulent bronchitis. The other unusual feature they
noted was the 'consolidation' of tissue near to the roots of the
lungs. The medical researchers noted that the consolidation

occurred where the infection had progressed to bronchopneumonia and was often associated with a localized inflammation of the lung tissue. The most alarming aspect of their investigation, however, was the disease's resistance to treatment. They tried everything they could think of – oxygen therapy, steam inhalation, even venesection (blood letting) – all to no avail. The only hope was passive immunization with a blood serum vaccine but, they concluded somewhat gloomily, 'it is unlikely, in view of the blocked condition of the bronchioles [to be] of great benefit.'

Etaples was not the only place where Army medics reported peculiar respiratory outbreaks that winter. In March 1917 a very similar disease was observed at an Army barracks in Aldershot by a young RAMC Major, Adolphe Abrahams.[43] The older brother of the Olympic sprinter Harold Abrahams (famously portrayed in the film *Chariots of Fire*), Abrahams would go on to enjoy a distinguished career as a pioneer of sports medicine, first as an honorary medical officer to the British Olympic athletic team and later as the founder of the British Association of Sport and Medicine. However, in the winters of 1916 and 1917 he'd been physician-in-charge of the Connaught Hospital at Aldershot barracks when a series of patients had presented with 'peculiar' lung infections. At first, Abrahams thought the cases were ordinary bronchopneumonia but he soon noticed that the patients expectorated large quantities of 'almost pure pus' rather than the 'frothy' expectoration that was usual in bronchitis. Moreover, although the patients' rapid breathing and elevated pulse rates suggested pneumonia, on closer examination the wheezing was confined to the base of the lungs rather than being widespread. The most distinctive feature of the illness, however, was 'a peculiar dusky heliotrope type of cyanosis of the face, lips and ears, so characteristic as to hall-mark the nature of the patient's malady even on superficial inspection.' Although some patients who reached this stage recovered, the cyanosis usually indicated an 'extremely bad' prognosis – the disease was resistant to every form of treatment and had a very high fatality rate, with between a quarter to half of the infections resulting in death.[44]

Significantly, Abrahams and his colleagues at Aldershot reached their conclusions quite independently of the Etaples researchers, and only realized they were dealing with the same disease when Hammond's *Lancet* paper was already well advanced. Writing to the *Lancet* in the summer and autumn of 1917, Abrahams expressed the view that it was probable that purulent bronchitis had been more widespread than he or other researchers had realized at the time. He also predicted it would probably reappear in the winter months – hence the importance of developing some sort of prophylaxis.

***

Army medical officers were not the only ones who found the conditions at Etaples disturbing. Long before scientists posited a connection between Etaples and the emergence of the influenza virus, this immense congregation of men and animals within the sound of shellfire had made the camp a surreal and, to some young imaginations, dreadful place. One person who felt distinctly uneasy on arriving at Etaples was Wilfred Owen. In January 1917, Owen, wrote to his mother to inform her that he was en route to Etaples where he was looking forward to joining the Second Manchesters, a regiment of well-trained troops and, as Owen put it, 'real-old Officers.'[45] But no sooner had he reached the camp than his enthusiasm evaporated. As he recalled in a letter to Susan Owen 11 months later, Etaples 'seemed neither France nor England, but a kind of paddock where the beasts are kept a few days before the shambles.' From his 'windy' tent in the middle of this 'vast, dreadful encampment' Owen could hear the revelling of Scotch troops – men who, he reflected, were now dead and 'who knew they would be dead.' But chiefly Owen thought of the strange look he'd seen on the faces of the Tommies. It was, he wrote, 'an incomprehensible look, which a man will never see in England, though wars should be in England; nor can it be seen in any battle. But only in Etaples. It was not despair, or terror, it was more terrible than terror, for it was a blindfold look, and without expression, like a dead rabbit's.'[46]

Vera Brittain, who spent ten months at Etaples, was similarly impressed by Etaples, writing that she remembered her time at the camp 'as vividly as anything that happened in my various hospitals.' An ardent supporter of women's suffrage, Brittain had joined the VAD because she wanted to make an active contribution to the war effort. But unlike other women of her class and background she was determined to be more than a mere auxiliary and had honed her nursing skills during gruelling service in a military hospital in Malta in 1916 – service that had earned her a scarlet braid known as an 'efficiency stripe'. The result was that when her brother Edward was sent to the Ypres Salient in August 1917, Brittain applied to be transferred to Etaples, calculating that at the camp, just 50 miles from the Front, they might have more opportunities for meeting.[47]

Brittain had little difficulty persuading her superiors. Etaples was in desperate need of skilled casualty nurses and her transfer was approved immediately. But, ironically, she had more contact with German soldiers than British and managed to meet up with Edward just once before he was killed in action. The visit took place in London from where Edward had telegraphed in January 1918 to say he'd been unexpectedly granted leave. Brittain, who by now had been working at

Etaples without a break for five months, immediately caught a ferry from Boulogne only for her reunion with her brother to be very nearly ruined when she developed a fever en route to Victoria, forcing her to spend the next ten days in bed.

Was this disease also a precursor of the pandemic strain? Unfortunately, from the meagre records available, it is impossible to say. Brittain writes only that the '"bug"... was difficult to locate' but was 'obviously a form of "P.U.O" or trench fever.' Whatever the case, Brittain almost certainly contracted it on the acute medical ward of No. 24 General Hospital, the same ward where Underdown had died the previous February. Brittain had found herself on the ward during the bitterly cold winter of 1917–18 and, although there is no evidence that purulent bronchitis was still raging then, Brittain describes treating several patients with gas gangrene, an infection caused by *Clostridium* bacteria, and severe pneumonia. The gas-gangrene cases were 'doomed from the start,' she writes, but far worse were the screams and mad rantings of the patients dying from pneumonia. One evening Brittain had to call two orderlies to assist her because she was being chased round the ward by 'a stark naked six-foot-four New Zealander in the fighting stage of delirium.' No sooner had the staff strapped this 'insane giant' to his bed than another nurse had to give another man dying of pneumonia a calming injection to stop him repeating a refrain about wanting to go to Trouville.

Such cases 'were a disturbing contrast to the sane, courageous surgicals,' Brittain reflected. 'Wounded men kept their personalities even after a serious operation, whereas those of the sick became so quickly impaired; the tiny, virulent microbe that attacked the body seemed to dominate the spirit as well.'

Brittain had no explanation for why personality was so vulnerable to these 'small, humiliating assailants.' Nor could she forget the 'heaps of gas cases' who arrived in the middle of the night, 'burned and blistered all over' and with their blind eyes 'stuck together, and always fighting for breath... saying their throats are closing and they know they will choke.' Such experiences – coupled with her time on the 'German ward' where she'd held the hand of an emaciated German officer even as Edward was trying to kill men just like him a few miles away – convinced Brittain of the futility and folly of war. 'The world was mad, and we were all victims; that was the only way to look at it,' she writes.

But Brittain's most enduring memory of Etaples that winter was not of the British and German soldiers lying near one another but of the bitterly cold, inhuman conditions:

Whenever I think of the War to-day, it is not summer but always as winter; always cold and darkness and discomfort... Its permanent symbol, for me, is a candle stuck in the neck of a bottle, the tiny flame flickering in an ice-cold draught, yet creating a miniature illusion of light against an opaque infinity of blackness.

PART II

'Don't cry-ee, don't sigh-ee,
There's a silver lining in the sky-ee,
Bonsoir, old thing, Cheer-i-o, chin, chin
Nah-poo, take the flue and die-ee.'

– corruption of popular wartime song

3

first wave, March–August 1918

In 1890, when Winston Churchill was a 15-year-old pupil at Harrow, the future British Prime Minister wrote a curious poem describing the recent progress of the Russian flu across Europe. To this precocious and patriotic young schoolboy, the flu had seemed a 'vile, insatiate scourge' – a plague which England had been lucky to escape. Tracing the virus's 'noiseless tread' from China and over 'bleak Siberia's plains' to Russia, Alsace and 'forlorn Lorraine,' Churchill wrote:

> The rich, the poor, the high, the low
> Alike the various symptoms know
> Alike before it droop[48]

Churchill's precocious verse captures perfectly a young man's dread of this exotic and mysterious disease. To Churchill's insular way of thinking it was only because of a quirk of geography – 'the streak of brine,' as he puts it – that the virus's power 'to threaten Freedom's isle' had been dissipated. Yet, as we have seen, if Britain suffered relatively few casualties from the first wave of Russian flu in 1890, this was certainly not true of the second wave that fell on Sheffield and other northern cities in the spring of 1891. Moreover, in 1918–19 this pattern of a mild primary wave followed by a severe secondary 'killing' wave would be repeated. Unfortunately, by 1918, the Russian flu was a distant memory – an irrelevance even for those who like Churchill had once been moved to poetry by its emblematic power – and Britain had far more pressing concerns. The result was that the insights into the epidemiology of flu gleaned from the Russian pandemic were ignored. The consequences would ultimately prove catastrophic.

In the winter of 1917–1918 the issue uppermost in most Britons' minds was not influenza but war and the truth was that for all that Beaverbrook's *Express* and Northcliffe's *Mail* praised the courage and heroism of the British Tommy, and for all the advertisements reminding civilians of the importance of 'staying power,' most Britons were weary of it. True, the news the previous April that the United States had formally joined the Allied cause had briefly boosted morale on the Home Front but the US had been slow to mobilize and General Pershing's promised reinforcements were still a long way from Bordeaux. Worse, in December Russia's new Bolshevik government had signed an armistice with the Central Powers, enabling Germany to move troops from the east to reinforce its positions along the Hindenburg Line. As Caroline Playne, a young historian living in Hampstead, north London, recorded in her diary the day after the Bolshevik announcement: 'There is an all-pervading atmosphere of dread…Great depression on most people's faces.'[49]

The First World War was a 'total war' in a sense that it is difficult for us to conceive of today. Although it had begun as a confrontation between professional armies, it had quickly degenerated into a war of attrition, a slow but deadly fight to the finish requiring the full mobilization of the civilian economy and civilian manpower. By 1916, with the introduction of conscription, the war had also become a matter of life and death for tens of thousands of conscripts – the majority of them young working class men recruited from Britain's industrial heartlands – and by the summer there were few families that had not been touched by the carnage. Rudyard Kipling, who lost his son Jack at the Battle of Loos, and who would later pen the bitter epitaph, 'If any question why we died/Tell them, because our fathers lied,' is perhaps the best-known example.[50] But the mourning of individual fathers and mothers pales into comparison with the suffering of entire communities following the Battle of the Somme – a campaign in which the British lost 57,000 men on the first day alone and which hit working class communities in West Yorkshire and Tyneside particularly hard.

The battle, which began in July and lasted five months at a cost of 485,000 British and French casualties, should have marked a Rubicon – Lloyd George called it 'the most gigantic, tenacious, grim, futile and bloody fights every waged in the history of the war.' But the following year the Somme was supplanted by another wasteful and equally pointless confrontation.

The Allied assault on the village of Passchendaele overlooking the water-logged Ypres salient lasted from July to November 1917 and has become a by-word for military miscalculation and hubris. General Sir Douglas Haig, the commander of the British Expeditionary Force (BEF) was confident he could take Passchendaele and the nearby Gheluvelt ridge in a matter of hours with simultaneous attacks on the German positions. Instead, British troops became bogged down in the Gheluvelt Plateau, floundering in the Flanders mud where they became sitting ducks for the German gunners. As the autumn rains turned the salient into a soupy mire, thousands of men died from a combination of wounds and the diseases that festered in the sludge. Even after the Canadians finally took Passchendaele on 6 November there was little sense of accomplishment. In all, the British had lost a quarter of a million men, and for what? As Haig's critics pointed out, no German communication lines had been cut and the Army's morale was in tatters. Moreover, when Haig renewed his assault on the Germans at Cambrai a few weeks later, briefly breaking through along a 10,000-yard front with the help of 300 British tanks – the first to be deployed in a major battle on the Western Front – he was so short of infantry he was unable to press home the advantage. The German defences were overrun to a depth of four miles prompting the ringing of church bells – the first to be sounded in England since the declaration of war three years earlier. But the celebrations were premature and when on 30 November the Germans counter-attacked, recapturing more ground than they had lost, the bells fell silent.

'The German papers are laughing over the premature joy bells of St Paul's,' recorded a retired British Army officer in his journal. 'It is said that only 4 tanks remain of those engaged.'[51]

If by the end of 1917 England had grown weary of war and Haig's repeated predictions of victory, it had also grown weary of rationing and other civilian restrictions deemed necessary for the successful prosecution of the war. The week before Christmas *The Times* reported seeing women, 'some with infants in their arms and others with children at their skirts,' patiently queuing outside dairy shops at 5am on a Saturday. In Broad Street, in the heart of the City, the lines for margarine were 1,000-deep, while in Walworth Road, southeast London, the paper estimated 3,000 people had queued for provisions, only for a third to be sent away empty handed two hours later.[52] By January, following the imposition of sugar rationing in London, there were riots in Leytonstone as munitions workers, whose wives were unable to secure essential foodstuffs, looted closed shops forcing the authorities to despatch the Army to quell the demonstrators, and in Manchester 100,000 workers marched on the Town Hall.

These protests were not fuelled so much by the shortages – which though inconvenient were regarded by most Britons as necessary – as by the fact that entrepreneurs seemed to be profiting from the misery. This resentment against 'war profiteering' was ameliorated somewhat in April 1918 when the government extended rationing, which now included meats and fats, nationwide. Even so, for munitions workers grown wealthy on overtime payments, or for those simply born to wealth, there were still ways around the queues, as the leading London department stores' lucrative trade in luxury grocery hampers and 'club' dinners for officers attested.

The same could not be said of those foolhardy enough to venture out on the dark but fine nights favoured by the German dirigibles. Death by Zeppelin or Gotha bomber could strike at any time, regardless of wealth or breeding. On the night of 28 January 1917 a Zeppelin attack killed 58 people in the capital and injured a further 173. In another incident a bomb exploded in the cellar of a printing works igniting a fire that killed 38 and injured 85. The raids, which also hit the Midlands and towns on England's exposed east coast, gave new meaning to the phrase 'war on the Home Front' and soon Londoners became used to listening for the burst of the maroons, a type of firework that signalled that an air raid was imminent. However, for all that the German airships had introduced a new source of anxiety into civilian life the majority of raids were unsuccessful and by May, with the Germans launching a make-or-break offensive on Allied positions in northern France, the attacks petered out. Influenza, on the other hand, was only just getting started.

Following the waves of Russian flu in the early 1890s influenza had become such a familiar feature of the winter 'catarrh' season that some medical authorities considered the disease almost endemic to Britain. In 1895, 1905, 1908 and 1915 there had been upswings in the incidence of flu accompanied by marked increases in the death rate. But while the winter was recognized as a dangerous season for the disease, severe outbreaks in the spring and summer were rare. And for all that British medical journals complained that influenza was capricious and still poorly understood, the majority of Britons probably shared the scepticism of Eliza Doolittle, the heroine of George Bernard Shaw's 1913 play *Pygmalion*, who seems to have regarded influenza as a highly suspect diagnosis.[53] Indeed, Britons' major preoccupation in the latter half of 1917 and early 1918 was with putting sufficient calories on the table. Although U-boat attacks on north American shipping in the winter of 1916–17 had resulted in severe wheat and bread shortages, shipments of grain had continued to slip through the German submarine blockade. However, with the mobilization of upwards of one million US troops in the summer of 1917, the surplus from American farms was now needed for the US Army. The result was that 'government bread,' the staple of most working class British diets, was of increasingly poor quality and the substitution of inferior grains and potato flour commonplace. Consumption of fresh meat and butter also decreased and with the introduction of rationing sugar consumption fell to something like half the pre-war level. Having said that, demographic studies show that Londoners enjoyed higher average calorific intakes during the war than either their Berlin or Parisian counterparts and that for those who survived the fighting post-war life expectancy actually increased.[54] Yes, there were still bottle-necks in the supply of tea, butter and margarine but as Lloyd George informed a deputation of trade union leaders to Downing Street shortly before Christmas 1917 he was pressing America for extra food credits, and despite four years of fighting there had been no famine. On the contrary, there was probably 'less hunger and privation' than had existed before the war, the Prime Minister argued. The important thing was to make sure the distribution of food was fair. The community needed to feel that as far as possible there was an 'equal sacrifice,' he said.[55]

***

No one can be sure where the virus that triggered the first wave of the 1918 pandemic came from but the earliest and most authoritative report comes from Haskell County, Kansas. In the early spring of 1918 Fort Riley in Haskell County was the second largest military cantonment in the US.[56] Built at the confluence of the Smoky Hill and Republican Rivers and surrounded by rich farmland, Fort Riley was originally designed as a cavalry barracks – 50 years earlier, during the Indian Wars, General Custer had set out from Riley in an attempt to sub-jugate the Cheyenne and the Sioux. However, following the US's decision in the spring of 1917 to declare war on Germany, an army encampment had been hurriedly constructed within the sprawling Fort Riley military reserve and by March 1918 Camp Funston held some 56,000 troops – the majority of them raw Midwestern recruits who had never been further east than St Louis. It was here that on 4 March Camp Funston's cook, Albert Gitchell, reported sick with a sore throat, fever and headache. Today, many scientists believe that Gitchell was the Great Flu's patient zero.

Normally, Gitchell would have been up before dawn to prepare breakfast but when he awoke on the morning of 4 March after a sleepless and restless night he felt so unwell he reported to the camp infirmary. There, the duty medical officer took Gitchell's temperature and, alarmed to find he was running a fever of 40° C, ordered him to an isolation bed as a precaution.

The next man to report to the infirmary was Corporal Lee W. Drake. A truck driver assigned to the Headquarters

Transportation Detachment's First Battalion, Drake had almost identical symptoms to Gitchell, so naturally the medic sent him to the same isolation ward. Then the medic turned to the next man in line, Sergeant Adolph Hurby. Hurby's temperature was even higher than Gitchell's – 41° C. Not only that but Hurby's throat, nasal passages and bronchial tubes were inflamed and when the medic shined a light in his eyes he winced with pain. Two men with identical symptoms might have been dismissed as bad luck but three was too much of a coincidence to ignore.

The medic picked up the phone and called his boss who in turn alerted Fort Riley's chief medical officer, Colonel Edward R. Schreiner. A 45-year-old army surgeon, Shreiner presided over the 3,068-bed hospital and had long feared an outbreak of respiratory disease in the camp's overcrowded and hard-to-heat barracks. Schreiner was right to be concerned: by noon bemused medical staff had 107 cases on their hands, and by the week's end they had 522. The following month a total of 1,100 men were incapacitated forcing Schreiner to requisition a hangar to supplement the overflow from the hospital.

At first glance, the illness resembled classic influenza: a chill followed by a high fever, headache and back pain. But some patients were so ill they found it impossible to stand up, earning the disease the nickname 'knock-me-down fever.' There was another disturbing feature too: while most men recovered in three to five days, 48 had died. Schreiner recorded the cause as pneumonia, but these were no ordinary pneumonias. In many cases the symptoms were alarming, horrifying even: violent coughing, projectile nosebleeds, and a deathly blue discolouration of the face. Schreiner took several cultures and finding Pfieffer's bacillus, *H. influenzae*, telegraphed Washington. His message, dated 30 March 1918, read simply:

'Many deaths influenza following immediately two extremely severe dust storms.'[57]

In fact, the weather almost certainly had nothing to do with it. Influenza is a communicable disease spread by birds, pigs and horses, not by the wind. If the pandemic strain did start at Camp Funston, it is far more likely that it originated on one of the nearby farms, where pigs, poultry and other livestock crowded together. Indeed, on 30 March – the same day that Shreiner had telegraphed Washington – *Public Health Reports*, the weekly journal of the US Public Health Service, carried a report from a local Haskell County practitioner warning of 'influenza of a severe type.' The disease had broken out on several local farms in January and February, overwhelming his practice before disappearing just as suddenly in mid-March. Interestingly, several of the victims had been soldiers en route to Camp Funston.

Was Haskell County the epicentre of the first wave, or could the same virus have been circulating at Etaples and other British Army camps at an earlier date? And if so, how could the virus have travelled east across the Atlantic to Kansas when most of the human traffic that spring was in the opposite direction? These questions are discussed in more detail in chapter six. For present purposes, it is sufficient to note that the report in *Public Health Reports* was the first official reference anywhere in the world to what may have been the pandemic strain of influenza.[58]

At first the medical authorities in Washington were not overly alarmed: after all, Camp Funston was just one army camp among many. However, when similar outbreaks were reported at a nearby school for Indians in Haskell County followed by US Army camps in Georgia and South Carolina, they grew increasingly worried. Then, in April, a dozen more camps were hit including Camp Dix, New Jersey. By now mini-epidemics were also being reported at other military camps along America's eastern seaboard, as well as in cities as far afield as Detroit and Chicago. More worryingly, from the Allied point of view, influenza had also begun to infect the American Expeditionary Force (AEF). Some 84,000 'doughboys' crossed the Atlantic on transports in March, followed by a further 118,000 in April. The arrival of these cheerful, well-fed Mid-Westerners came as a tonic to those who, like Vera Brittain, had struggled through the grim winter of 1917. Emerging from the ward at Etaples where she'd nursed delirious pneumonia cases the previous December, Brittain describes stepping out into the spring sunshine as the first divisions of Americans marched through the camp. Compared to the 'tired, fever-racked' British troops, the Americans struck her as 'god-like' and 'magnificent.' She writes: 'There seemed to be hundreds of them and in the fear-less swagger of their proud strength they looked a formidable bulwark against the peril looming from Amiens.'[59]

Brittain was referring to the fact that the German Army under its commander, General Erich von Ludendorff, was now within 12 miles of the key rail centre. The devastating German attack on Amiens – part of Ludendorff's long-threatened 'spring offensive' – had begun on 21 March when 200 German divisions, backed by specialized infantry assault units equipped with light machine guns, mortars and flame throwers, had begun advancing on a 40-mile front. Haig appealed to his opposite number for reinforcements but the French Commander-in-Chief Henri Pétain refused for fear of exposing his rear and Paris to attack. The War Cabinet were informed and an emissary was hurriedly despatched to Versailles to alert the French president Georges Clemenceau to the impending disaster. As a result the French agreed to install the French Chief of Staff Marshall Foch as the overall Allied Commander and coordination between the French and British was restored. However, by now Ludendorff had switched his point of attack to the north in an attempt to seize the Channel ports. One by one, the Allied positions won at such cost only months before were abandoned and on 9 April the Germans had recaptured all the ground west of Ypres that it had taken the British three months and 400,000 men to conquer the previous autumn. On 11 April, with the Germans poised to retake Passchendaele, Haig issued his famous 'backs to the wall' Order of the Day. A strategic decision was taken to evacuate Passchendaele and following a further shortening of the line between Armentieres and Ypres the German advance was halted. But Ludendorff was not finished. Instead, he began drawing up plans for one more bold assault – this time against the French positions guarding the Aisne, a line thought to be impregnable by the Allies. By now American troops were arriving in France at the rate of 150,000 a month and Ludendorff knew that time was running out. In an attempt to rally his weary troops he told them the offensive would be a *Friedenssturm* – a blow for peace. What Ludendorff couldn't know was that both the Germans and the Allies now faced an invisible and deadly new enemy.

***

In the middle of April there was an outbreak of influenza at Brest, the chief disembarkation port for American troops, followed a few days later by further outbreaks at AEF camps in the Marne and Vosges. 'Epidemic of acute infectious fever, nature unknown,' reported a medical officer at an army hospital in Bordeaux on 15 April. The symptoms were identical to those at Camp Funston – severe headache, backache and chills, followed by prostration, high fever and, in a few cases, inflammation of the upper respiratory tract. But, as at Funston, the duration of the disease was short and the symptoms relatively mild, prompting AEF medics and officers to refer to it as 'three-day fever.' By May, with thousands of French soldiers also sick with chills and fever, American troops began openly referring to the disease as 'la grippe.' Then, towards the end of the month, came the news that 'a strange form of disease of epidemic character' had broken out in Madrid forcing the city to close its theatres and curtail its tram services. In other parts of Spain up to a third of the population – or some eight million Spaniards – were similarly stricken. But the greatest public concern was reserved for the Spanish King, Alfonso XIII, who was said to have fallen ill while attending a religious service in the palace chapel. As the illness spread to other members of his government, including the Prime Minister and Finance Minister, Madrid wits started referring to the disease as 'the soldier of Naples,' an appellation taken from the lyric of a popular operetta.[60] In the reports carried by British and other foreign newspapers, however, the disease quickly acquired the sobriquet 'Spanish' influenza. The name stuck not because Spain was the first country to be visited by influenza in the spring or because it was the only one where the flu was extensive. Rather, it was because newspaper proprietors in Britain and other belligerent countries were reluctant to give publicity to local flu outbreaks for fear of provoking a 'civilian scare'. By contrast in Spain, which was not party to the war, there were no such concerns and both domestic and foreign correspondents freely documented influenza's depredations there.[61]

Whatever its origins, by 2 June *The Times* was reporting that the flu was responsible for 700 deaths in Madrid and the epidemic had 'passed the joking stage.' From Spain, the disease spread throughout the Iberian peninsula. In May it also invaded Greece, Macedonia and Egypt, and in June sporadic outbreaks were reported in Germany, most likely as a result of soldiers returning from the Front. In July, flu was reported in Denmark and Norway, but in Holland and Sweden the disease did not arrive until the following month.

In far-flung European colonial outposts there were further reasons to doubt the theory of a Spanish origin. In India, the disease was first reported on a transport at Bombay on 29 May from where it spread via the railways to Calcutta, Madras and Rangoon. But if the Indian disease was truly 'Spanish' influenza it would have had to have flown – in 1918 it took a week to ten days to reach India by boat or train. At around the same time influenza was reported in Shanghai and there were also outbreaks in the Japanese Navy. In Switzerland reported cases of influenza jumped from a mere six in June to nearly 54,000 in July. By August the disease had reached Peru, and the following month it arrived in New Zealand and Australia. Whether from a single vessel or multiple vessels, the influenza pot appeared to be boiling over.

As the virus jumped from one army camp and battalion to another, the generals became increasingly concerned. Towards the end of June, the Director General of the British Army Medical Services commandeered ambulance trains to transport sick men direct from their units in the First Army to Etaples only to see a hospital ward there fill up with 3,000 men in three days. The disease would seize soldiers so suddenly and 'with such a sense of prostration,' recorded one senior Army doctor that they would be 'utterly unable to carry on.' His account continues: 'From sheer lassitude [the patient] would be obliged to lie down where he was, or crawl with difficulty back to bed so that barrack rooms which the day before had been full of bustle and life, would now be converted wholesale into one great sick room.'[62]

By now the AEF was also reporting phenomenally high sick rates – as much as 80 or 90 per cent of a battalion's fighting strength. Then came the news that the Grand Fleet at Scapa

Flow had been affected. Churchill's 'insatiate scourge' was on the march and this time neither the 'streak of brine' nor the wide-open waters of the North Sea appeared able to retard it.

The only consolation was that, if anything, the flu was hitting the Germans even harder. On 27 May Ludendorff had renewed his offensive, launching three armies across the Aisne towards Paris. The attack took the worn-out British and French divisions completely by surprise and within five days German storm troops had reached the banks of the Marne. Ludendorff was now just 40 miles from Paris. But just as it looked as if he would breach the French lines and take Reims his advance ground to halt. This wasn't so much due to the spirited French and American counter-attacks as to the fact that by now Ludendorff's divisions were completely overstretched. There may have been another factor too: influenza. As Ludendorff would recall in his memoirs: 'It was a grievous business having to listen every morning to the chiefs of staffs' recital of the number of influenza cases, and their complaints about the weakness of their troops if the English attacked again.'[63]

News of the Allied successes came as a much-needed tonic on the Home Front. In March, when the news from northern France had been unremittingly gloomy, Caroline Playne had noted in her diary how 'people looked frightfully strained.' But by May, with the news that the British line at Ypres had held, she noted that Londoners had 'throw[n] off depression & became more cheerful than at any other time during war.'[64] As May gave way to early summer the news from the Front got steadily better. In late June, following a series of pitched battles around Noyon, the French and Americans halted the German drive at Compiègne and began to advance on Château Thierry. Then in the first week of July the Americans routed the Germans at Vaux and the French took more than 1,000 German troops prisoner at Autreche. In the middle of the month, Ludendorff launched simultaneous attacks across a 25-mile front between Champagne and the Marne in one last desperate attempt to breach the French lines and reach Paris. It would be his last roll of the dice. Although the German Seventh Army succeeded in crossing the Marne, the Allies halted the German advance to the east of Reims on the first day. Then, on 18 July the French and Americans smashed Ludendorff's flanks and, despite stubborn resistance, began pushing the Germans back towards Soissons and Chateau-Thierry.

'The whole temperature of London has gone up in these two days of good news,' noted the *Manchester Guardian's* London correspondent on 20 July. 'Although there will be no flags or bells for some time yet the look on the faces of the people is like flags and bells.'

The encouraging news from the Front, combined with the mild nature of the spring and early summer wave of flu and the impression, encouraged by the newspaper reporting, that influenza was largely a Spanish or, at least, a foreign problem, caused most Britons to view the disease's arrival with equanimity. As *The Times's* Glasgow's correspondent commented:

> The man in the street, having been taught by that *plagosus orbilius*, war, to take a keener interest in foreign affairs, discussed the news of the epidemic which spread with such surprising rapidity through Spain a few weeks ago, and cheerfully anticipated its arrival here.[65]

However, for army medics who had seen the flu sweep through the front lines at first hand and had taken note of the extensive morbidity, and for civilian doctors and public health officials who had made a special study of the disease, there were plenty of reasons to be concerned. The first was that although just 5,500 British troops had died from the flu – a paltry figure when set against the dreadful arithmetic of the

Somme and Passchendaele – some 226,000 had been hospitalized, a truly staggering number. Moreover, it was these same soldiers, returning to Britain via Portsmouth and other Channel ports, who had introduced the disease to the civilian population, carrying the virus by rail to London and Birmingham from where it was refracted by the train network to northern cities such as Leeds and Liverpool, and west to Bristol and Cardiff.

'In a typical case the onset is strikingly sudden,' observed Oliver Gotch, a Royal Naval surgeon temporarily based at the Central Royal Air Force Hospital, Hampstead, north London.

A patient may go to bed apparently well and wake up complaining of body pains, headache, malaise, etc.; or he may be at his work, when he suddenly feels violently giddy, and in a few seconds may fall to the ground in a state of collapse... The second day of the disease finds the patient a good deal worse, with, in addition, painful photophobia [fear of light] and there is a great desire to sleep the whole day... The third day will, as a rule, see an improvement in the general feeling of discomfort... convalescence is rather slow, and the majority of patients do not recover their strength for a week to ten days after getting up.[66]

The speed of the influenza attacks also took factories and works by surprise and by July few parts of the country were unaffected. In Lancashire, *The Times* reported, one textile house employing 400 workers had 100 people laid up. In Wigan a third of mineworkers were off sick and in Newcastle absentee rates in collieries were running at 70 per cent. Meanwhile in Manchester the medical officer of health, James Niven, had been so alarmed by the disease's impact on schoolchildren he'd immediately requested the closure of the city's schools. '[The children] simply dropped on their desk like a plant whose roots have poisoned, the attack being quite sudden, and drowsiness a prominent symptom,' he observed.[67] Niven had 35,000 handbills printed and distributed to local factories and businesses spelling out in clear and easy-to-understand English the dangers of influenza and giving strict instructions for the isolation of the sick. His prompt measures may have helped: on 18 July the city's education committee agreed to close all the schools and although flu attacked 100,000 Mancunians during the spring and early summer just 322 died, a relatively low mortality rate. However, while doing the rounds of schools and factories Niven had been alarmed by the way the disease seemed to single out the healthiest and fittest members of a workforce or community. His concern was informed by his experience of the 1890 pandemic when he'd been the medical health officer for Oldham.

Born in Peterhead and educated at Cambridge and St Thomas's Hospital, London, Niven was a shy and retiring man who lived for his work. Dedicated to public health, he ascribed to the adage that 'an ounce of prevention is worth a pound of cure.' Accordingly, when in 1890 the Russian flu had visited Oldham he issued strict orders for the disinfection of contaminated premises and the isolation of the sick. Niven also advised that those who had been infected should wait at least three weeks before returning to work to ensure they would not spread the contagion more widely. His measures were successful and when there was a recrudescence of the Russian flu in the spring of 1891, followed by a third wave in 1892, Oldham was not as badly affected as other northern British towns.

There was a key difference, however, between the Russian flu and the latest manifestation of the disease. In 1889–92 it was the elderly and the very young who'd proved most vulnerable – the same pattern as in normal winter flu season. But in the summer of 1918 the flu had sickened everyone. The only consolation was the large numbers of people who'd been attacked in the first wave. If the Spanish influenza behaved like the Russian flu, reasoned Niven, then those people should at least enjoy immunity when the second wave hit.

Niven wasn't the only medical man concerned by the summer wave. At the Medical Research Committee (MRC) in London the committee's energetic young secretary, Walter Morley Fletcher was also growing worried. A brilliant biochemist and former Cambridge blue, Fletcher was a tall, athletic man with a sharp intelligence and easy charm. A graduate of Trinity College, Cambridge, where he had obtained a double first in natural science, Fletcher had trained as a doctor at St Bartholomew's Hospital, London, before returning to his former college as a senior demonstrator in physiology. The main focus of his studies was the biochemistry of muscular contraction – research which eventually resulted in his election to the Royal Society – but while at Cambridge he had also befriended Sir Robert Morant, a civil servant who was the architect of Lloyd George's national insurance reforms. In 1914 Morant approached Fletcher to head up the MRC, a new body tasked with providing impartial scientific advice to government, thus advancing Sir Robert's public service goals of bettering the nation's health. But no sooner had Fletcher taken up his post with a budget of £50,000 a year than the war broke out.

Working closely with Sir William Boog Leishman, a tropical medicine specialist and Director General of the Army Medical Services, Fletcher was appointed to the Army Pathological Committee where he was thrust into the forefront of the battle against sepsis, gas-gangrene, and other 'diseases of war.' As a result, he increasingly found himself having to liaise with medics, pathologists and bacteriologists in the RAMC. Fletcher was the perfect man for the job as he had known many of the officers in their civilian garb and the MRC and Army medical service soon forged a close working relationship. However, by early 1916 the stress of juggling his military and civilian responsibilities proved too much and that winter Fletcher contracted pneumonia very nearly dying.

He returned to work just as the first wave of influenza arrived in northern France, sweeping through Etaples and other Army camps. The ground held by the BEF between the Channel ports and Ypres was dotted with stationary hospitals, each one equipped with its own laboratory staffed by an RAMC pathologist and bacteriologist. Most labs had very rudimentary equipment but at Abbeville there was a specialist unit where scientists could test for bacteria like Pfeiffer's bacillus, carry out immunological experiments with mice and other animals, and even manufacture rudimentary serum-based vaccines. The natural person to oversee this work was Fletcher and he soon found his desk piled with letters and reports describing the summer outbreaks of influenza and pneumonia and the various investigations into the pathology of the disease.

One of the more intriguing reports came from two RAMC officers based at a stationary hospital at Boulogne. In June and July the medics had examined the lungs of six soldiers who had recently died of bronchitis or pneumonia following attacks of influenza. Comparing their lung lesions with those from victims of gas attacks, they noticed a series of startling similarities. Although evidence of burning and necrosis was absent from the trachea of the influenza victims, their lungs displayed the same 'consolidation' of lung tissue seen in gas poisoning cases, as well as similar haemorrhaging of the bronchi and bronchioles. Speculating that the influenza 'bacillus' acted as a pioneer for other infectious bacteria, the researchers theorized that it was these bacteria that were responsible for the destruction of the lung tissue, concluding that 'poison gas would seem to do the same nefarious work.'[68]

At the same time as influenza was attracting the interest of bacteriologists, other scientists were collecting data on the incidence of the disease with a view to illuminating influenza's epidemiology and the degree to which one attack conferred immunity against another. One of them was Major Greenwood, a gifted young Army medical statistician who after the war went on to enjoy an illustrious career as professor of epidemiology at the London School of Hygiene and Tropical Medicine. The son of an East End doctor, Greenwood was educated at Merchant Taylor's school in the City and had gone to University College to study mathematics as a Buxton scholar. After graduating he completed his medical training at London Hospital, but in 1910 he decided to switch tack, joining the Lister Institute as a statistician. There he came under the sway of Karl Pearson whose *Grammar of Science* was to inspire a new generation of epidemiologists interested in the emerging discipline of biometrics. At the outbreak of the war, Greenwood enlisted as a captain in the RAMC and in 1917 he was put in charge of the medical research division of the Ministry of Munitions where he made a special study of fatigue and industrial wastage.

During his Army service Greenwood, like other medics, had become alarmed by the high battalion hospitalization rates and had decided to make a statistical study of influenza. Using Royal Air Force records Greenwood plotted a graph showing the increase in cases and compared them with the onset of Russian flu in 1889–90. Then, the first wave had occurred in winter, not summer, but Greenwood noticed the outbreaks exhibited near identical symmetrical 'curves' showing a rapid increase followed by a steep decline, like an inverted V. Although Greenwood would not be in a position to draw definitive conclusions until after the return of influenza in September, his worry was that the summer wave might be the herald of a more severe secondary wave. After all, this was what had happened during the Russian flu. The difference was that then the second wave had struck in the spring of 1891, a relatively clement time of year when people were less likely to be suffering from coughs and other respiratory infections. But Greenwood's fear was that in 1918 influenza would revisit the British Isles in autumn or early winter – the worst time of year for respiratory diseases when resistance was at its lowest.

It is not known if Greenwood and Fletcher corresponded on this point but it is clear from communications between Fletcher and other RAMC officers that Fletcher shared Greenwood's fears about an autumn recrudescence of influenza, explaining in November 1918 that 'it is natural to expect secondary waves with great confidence and as the primary wave came in the early summer, it was not a bad guess that a secondary wave with its dangerous pneumonia would come at the approach of winter.' Fletcher added that the MRC's approach had been to err on the side of caution 'and get ready for a secondary wave whether it came early or late.'[69]

However, for all that Fletcher may have been in a privileged position to alert senior Army and Whitehall officials, in reality there was little that either he or Greenwood could do. Sure, influenza had resulted in very high battalion sickness rates but within a week most men had returned to their units. The Army's main medical challenge was not influenza but sepsis and gangrene developing in open wounds and the lice and enteric fevers that bred in the trenches. Besides, in order to persuade Leishman and bureaucrats in the War Office to take action against influenza Fletcher would need hard scientific evidence and be able to put forward a definite plan of attack. But starved of researchers and independent facilities, Fletcher was not in a position to isolate the pathogen – whether it was Pfeiffer's bacillus or, as Fletcher had begun to suspect, a 'filter-passing virus' – let alone manufacture a vaccine against the dominant strain.

The problem was that in 1918 there was no single Whitehall department responsible for monitoring the nation's health and guarding against infectious disease. That would only come in 1919 with the establishment of the Ministry of Health. Instead, public health was the responsibility of a network of local sanitary authorities and local Medical Officers of Health (MOHs) appointed by town halls. In addition, the Board of Education carried out regular medical inspections of schools via local education authorities, and through the establishment of a series of National Health Commissions Lloyd George had recently begun the process of transferring responsibility for medical benefits from Approved Societies to new *ad hoc* bodies administered by local insurance committees. However, coverage was far from comprehensive and, once again, the quality of care varied from region to region.[70]

There was only one organization in Britain with the money and resources to mount a proper investigation of the summer influenza wave and that was the Local Government Board (LGB). But the LGB's main function was to administer the Poor Law and, except in the case of diseases that were the subject of specific parliamentary acts, it was reluctant to issue advice to MOHs or initiate independent investigations. The problem was that while the LGB had a statutory duty to collect information on infectious diseases such as cholera and tuberculosis, and communicable childhood diseases such as measles and whooping cough, in 1918 influenza was not a notifiable disease. Even if influenza had been notifiable, the LGB considered that there was little that the medical profession could do to treat or prevent it. During the Russian flu, for instance, the LGB had not seen fit to issue a memorandum until January 1892, nearly two years after the start of the outbreak. Then, it had advised sufferers to stay in bed, keep warm, drink brandy, take quinine and opium, and to have all bedlinen and clothing disinfected. True, there had also been recent epidemics of influenza in 1913 and 1915 but as Sir Arthur Newsholme, the Chief Medical Officer of the LGB, pointed out, each year bronchitis and pneumonia exacted a far heavier toll. If sanitary reforms had failed to make any difference to such entrenched respiratory diseases what chance did they have against influenza? Another option was quarantines but while cholera and plague could arguably be stayed by such restrictions, Newsholme pointed out that influenza travelled far more rapidly and was too extensively diffused. 'I know of no public health measures which can resist the progress of pandemic influenza,' he said.[71]

Still, the LGB could have initiated further bacteriological research with a view to developing a prophylactic vaccine. As we have seen, such research was already underway at various Army labs, albeit in a piecemeal fashion, and certainly by October Fletcher was urging Newsholme to be far more pro-active, writing to him to explain that the MRC was severely under-resourced due to so many bacteriologists being away at the Front and that he would be 'most grateful for any help you can give in investigations of the present epidemic.'[72] As the former MOH to Brighton – one of the worst affected towns during the Russian pandemic – and a person whose public service career had been devoted to the investigation and control of tuberculosis, Newsholme might have been expected to be sympathetic to such appeals. But though in July Newsholme began drafting plans to reduce absenteeism among munitions workers by laying on extra nursing staff, cut overcrowding on workmen's trains, and give police powers to close cinemas and places of entertainment, when in August Fletcher

Figure 3.1 Sir Arthur Newsholme
Credit: Wellcome Library, London

published a notice in the *BMJ* asking bacteriologists to forward their observations on influenza to the MRC Newsholme
failed to see it. Sometime that month he also shelved his
proposals for reducing the impact of influenza in the event of
its predicted return in the autumn.

Newsholme's reasons had little do with the preparedness of
the medical profession or medical knowledge about influenza
and everything to do with the war. 'On the balance of considerations,' Newsholme would later explain at a summit
meeting on influenza held at the Royal Society of Medicine on
13 November 1918, he had not considered it 'expedient' to
issue the memorandum. There were national circumstances,
he argued, 'in which the major duty is to "carry on," even
when risk to health and life is involved.'

> This duty has arisen as regards influenza among the belligerent
> forces, both our own and of the enemy, milder cases being treated
> in the lines; it has arisen among munition workers and other
> workers engaged in work of urgent national importance; it has
> arisen on a gigantic scale in connection with the transport during
> 1918 of many hundreds of thousands of troops to this country and
> to France from overseas. In each of the cases cited some lives might
> have been saved, spread of infection diminished, great suffering
> avoided, if the known sick could have been isolated from the
> healthy... But it was necessary to "carry on," and the relentless
> needs of warfare justified incurring this risk of spreading infection
> and the associated creation of a more virulent type of disease or of
> mixed diseases.[73]

Newsholme added that similar considerations prohibited the
LGB from issuing instructions to reduce overcrowding on
trains, trams and omnibuses. 'These doubtless are prolific
sources of infection, but the service cannot immediately be

increased, and meanwhile the vast army of workers must not be impeded by regulations as to overcrowding of vehicles in their efforts to go to work and to return home.'

With the benefit of hindsight, Newsholme's statement, delivered two days after the Armistice, reads like a belated attempt to justify his earlier inaction. Given what was at stake, it is hard to believe that he could have taken such a momentous decision in isolation. Having said that, there is no evidence to suggest that pressure was brought to bear on him by senior Army or Whitehall officials, and if Newsholme's statement reads like a belated apology we should remember that it was delivered *after* the recrudescence of influenza in the autumn when the disease had cut a swathe through British towns and cities killing more than 4,000 people in London alone. But in the summer when Newsholme was weighing up what to do he could not be certain that influenza would return in the autumn. Nor could he have envisaged that when it did the consequences would prove so devastating.

Newsholme was not alone. Newspapers and periodicals were full of advertisements for food and health supplements that affected similar disdain for influenza and other microscopic pathogens. '"Carry on" aptly describes the spirit of the nation to-day,' read an advert for the milk supplement, Bynogen. 'Endurance is the watchword of the allies. One of the great factors upon which endurance depends – that stable strength we call "staying power."'

Perhaps, it was this determination to carry on, coupled with the nation's naïve faith in the fortifying power of health supplements, that explains the behaviour of the writer Robert Graves's mother-in-law. In early July, like hundreds of thousands of other Londoners, she contracted influenza and was advised by her doctor to stay in bed. But when she learnt that her beloved son, Tony, had been granted leave from his unit and was due to pay her an unscheduled visit she hauled herself out of her sickbed and accompanied him on the rounds of the latest London plays. Graves reports that she died on July 13, 'her chief feeling [being]... one of pleasure that Tony had got his leave prolonged on her account.' What such recollections make clear is that in the summer of 1918 few people realized the danger posed by influenza and, in any case, the war and duty to family took precedence. For those who studied the runes, however, there were no shortage of signs.

On 24 June *The Times's* Dublin correspondent reported that outbreaks of influenza in the north and west of Ireland had sparked panic in working class districts and that in Belfast many schools and factories were closed. In Glasgow and Sheffield – northern cities that usually prided themselves on their stoicism – influenza was provoking similar alarm. In May three deaths were reported on ships moored in Glasgow harbour. From there the virus spread to Clydeside and nearby slum areas in the Govan and the Gorbals sparking a citywide epidemic. The Glasgow tenements were notoriously unsanitary – a breeding ground for typhoid and tuberculosis. Nevertheless, the eight-week flu epidemic appeared to have introduced a new threat: pneumonia. 'High death rate in Glasgow,' reported the *Glasgow Herald* on 17 July detailing 13 deaths from influenza and a further 26 from pneumonia. A week later, the death toll was even higher: 14 from influenza and 49 from pneumonia.[74]

The situation appeared little better in Sheffield where flu had begun to deplete the ranks of a workforce already stretched to the limit by conscription. At one works, the *Yorkshire Telegraph* reported, 15 per cent of employees were off sick and chemists were reporting a 'phenomenal' run on quinine. As casualties mounted and bodies began to pile up at local morgues, the Sheffield deputy town clerk appealed to the National Service Board representative to release men for grave-digging duty.

'People are lying dead in their houses seven days, and sometimes nine,' he told an appeals tribunal. 'The position is very serious indeed.'[75] Many local people blamed the disease on soldiers visiting Sheffield on furlough. As a result, the authorities decided to declare cinemas and other places of entertainment out of bounds to military personnel. Then towards the end of July came news of a serious outbreak of influenza at a German prisoner-of-war camp in Bramley, Hampshire. Nearly 1,000 men – a third of the camp's prison population – had been struck down forcing the authorities to convey the men to nearby civilian hospitals. Several guards were also now suffering from the same cause.[76]

The report, which made several daily and regional newspapers, prompted a subtle shift in the language used by journalists. Suddenly the 'Hun' was no longer the only enemy. 'New Foe in Our Midsts,' declared the *Salford Reporter*. 'The epidemic of influenza has reached Salford and if it is not of the old "sneezing variety," it is very prostrating. Hundreds of cases have occurred in the borough during the week, and doctors are extraordinarily busy.' The article went on to advise that victims should not attempt to carry on working but should go to bed the moment they were attacked. 'If you get about and try to shake it off it becomes much worse,' a doctor was quoted as telling the paper.[77]

But while some quarters of the press were happy to stoke fears of influenza, the most consistent advice was to ignore the disease lest fear itself open the way to infection. 'Malnutrition and the general weakening of nerve-power known as war-weariness provide the necessary conditions for an epidemic,' opined *The Times*. '...contact between national armies, which tends to make diseases international, is another factor favourable to propagation.'[78]

At the time this notion of flu as *sui generis* – a product of the stress and strains of war – struck most commentators as self-evident. Certainly, it seemed to explain why cities like Sheffield, Manchester and London, where there was a constant traffic of soldiers returning from the Front and where thousands of young women were employed in stressful munitions work, were particularly susceptible. However, while the stress and strains of war may have contributed to the high morbidity from flu, particularly in poor, overcrowded inner-city areas, the disease was no respecter of social class and was democratic in its choice of victims. Indeed, when influenza returned in the autumn, it would be precisely the fittest and healthiest members of the British population who would be hardest hit, suggesting that the critical factor was not the war but adaptive mutations in the virus itself or else an unusual interaction between the virus and the human immune system.[79] Finally, one should keep in mind that most munitions factories had their own canteens and that thanks to the plentiful opportunities for overtime most people employed in munitions work saw their incomes rise during the war.

But these were questions for future generations. For the moment, the pressing issue was the war itself. Faced with an Allied counter-attack on Château-Thierry Ludendorff had ordered his troops to pull out of the salient and to form a new defensive line along the line of the Aisne and Velsne rivers. Although the Allies succeeded in liberating Soissons on 2 August, the Americans could not evict the Germans from their new positions and the line held. Undeterred, the Allies decided to probe the salient which stretched across the old Somme battlefield near Amiens. This time nothing was left to chance. Selecting Sir Henry Rawlinson and the British Fourth Army to lead the attack, Haig moved every available tank into Rawlinson's sector. He also put extra men at Rawlinson's disposal, giving him a numerical superiority over the Germans of six to one.

Launched on 8 August the Amiens offensive was a stunning success. Rawlinson quickly broke through the German lines, routing six divisions and pushing the Germans back nine miles in one day – a defeat that prompted Ludendorff to remark that 8 August was a 'black day' for the German Army. Weakened by desertions – some 32,000 German soldiers fled the battlefield between July and August – the Germans had now retreated almost to the Hindenburg Line, a series of defensive zones they'd constructed over the winter of 1916–17 which lay just behind the Somme. Foch, the Allied-commander in chief, was eager for Haig to press home the advantage but Haig had learnt to be cautious and, fearing the strength of the German defences and worried that his tanks would become mired in the heavily scarred Somme battlefields, he decided to call a halt to the advance and search for more favourable ground.

The British Fourth Army was not the only conspicuous absentee. During the lull in the fighting influenza also disappeared from the battlefield and on August 10 the British command officially declared the epidemic over. Ten days later, an American medical correspondent writing from London confirmed that influenza had also deserted the British capital. 'The influenza epidemic described in recent letters has completely disappeared,' he declared.[80] In all, 700 Londoners had died from the flu and a further 475 from pneumonia during July. Ordinarily, those would be startling figures but Britain was still at war and no one, least of all Londoners, batted an eyelid.

4

second wave, September–December 1918

From his sick bed in Manchester town hall Lloyd George could just make out the tops of the plane trees lining Albert Square and, below them, the statue of John Bright dripping with constant rain. It had been raining for five long days now. The small group of well-wishers standing vigil outside the Prime Minister's bedroom had little doubt that the weather was the cause of his illness.

Lloyd George had first reported feeling unwell on the evening of Wednesday, 12 September, the day after he had arrived in Manchester to receive the freedom of the city. Although he had been raised in Wales from the age of one, Lloyd George had been born in Chorlton-on-Medlock and as he rode in an open carriage from London Road station the Manchester crowds had hailed him as a returning son. Indeed, the *Manchester Guardian* reported that along Piccadilly and Deansgate the crowds of cheering female munitions workers and soldiers home on furlough had been three deep and the 'turmoil in the streets' so great that it had taken Lloyd George nearly an hour to reach Albert Square.[81]

The following morning, the 'hero-worship' continued with more flag-waving and bell-ringing as Lloyd George made his way from the town hall to the Hippodrome for the presentation ceremony. Accepting the keys to the city from the Lord Mayor, Lloyd George quipped that the loss of his Mancunian citizenship was a circumstance over which he'd had 'no control.' He followed this with a rousing oration on the progress being made in the war, during which he pointedly praised the new unity of Allied command and argued that 'nothing but heart failure' could prevent Britain from defeating Germany. But though his speech had been greeted with loud applause, his aides remarked afterwards that he had 'not been in his best form' and later that evening, shortly before he was due to attend another function at the Reform Club, he collapsed.

The official line was that the Prime Minister had caught a 'chill' when on his return to the town hall he had lingered too long in the rain in Albert Square without a tophat and been soaked in successive downpours. 'It is sad that his native air should have proved thus insalubrious,' opined the *Manchester Guardian*, 'but he must forgive us for saying that he tempted its rigours too far.'[82]

In fact, Lloyd George had influenza – almost certainly the same strain that had swept through Spain, France and Britain in the spring and summer. Although mortality from the earlier wave had been low, Sir William Milligan, the eminent Manchester ear, nose and throat specialist who had been called to the Prime Minister's bedside, was not taking any chances. On his arrival at Manchester's neo-Gothic town hall on Thursday afternoon Sir William had immediately checked the Prime Minister's temperature and, finding that it was elevated and that he also had a sore throat, had ordered him to stay in bed and cancel his appointments. Mindful of the importance of maintaining national morale in time of war, the reports in the following day's papers downplayed the Prime Minister's illness, focusing instead on the reception he had received from the Manchester crowds and the message contained in his speech that as far as the fighting was concerned 'the worst was over'. The report in the following day's *Manchester Guardian* was typical, leading with the headline

'Peace without Chauvinism' and only mentioning in the third line that the Prime Minister had suffered 'a slight attack of illness after a long day' and that it was possible he might not be able to fulfil all of his scheduled engagements.[83] The reports were wildly optimistic. Indeed, when Sir William returned to the town hall on Friday morning he found the Prime Minister's temperature was still alarmingly high. By late afternoon there had been no visible improvement and at 6.15pm Sir William had had no choice but to issue a bulletin announcing that the Prime Minister was suffering 'an attack of influenza, accompanied by high temperature and complicated with a sore throat,' and would be unable to continue with his tour of Blackpool and other northern towns as planned.

No doubt Sir William thought long and hard before issuing such a specific diagnosis. Lloyd's George's illness came at a critical point in the war. The Prime Minister had been furious with Haig for recalling Rawlinson's tanks in early August just as they appeared poised to breach Ludendorff's defences on the Somme. Instead, Haig had switched his point of attack north, using General Byng's Third Army to advance on the German Second Army at Albert. The assault on 21 August achieved 'complete surprise' and the following day Haig had thrown caution to the wind, widening his line of advance and ordering his tank and infantry divisions to capture whatever ground they could regardless of the threat to their flanks. This time the Third and Fourth Army had pressed Ludendorff all the way, opening up a 40-mile front that stretched from Arras south to Bapaume and Peronne. During the Battle of the Somme, the Bapaume road had been the site of one of the bloodiest and costliest offensives in British military history, inspiring Siegfried Sassoon's satirical poem, *Blighters*, in which he lambasted the jingoistic music hall songs and jokes that, in his view, mocked 'the riddled corpses round Bapaume.' On 29 August, however, the majority of the corpses were on the German side. This time Bapaume proved a turning point, and following the Australian capture of Peronne two days later, Ludendorff was forced to retreat to the Hindenburg Line, abandoning all the territory he had won earlier in 1918.

The victory paved the way for the first independent American offensive of the war. The site chosen by General Pershing to blood his Midwestern troops was Saint-Mihiel. The salient, south of Verdun, had been in German hands since 1914, but on 12 September, with Ludendorff abandoning his positions, Pershing, backed by a French colonial division and 2,900 guns, launched a massive artillery barrage. Ludendorff was caught by surprise and within 36 hours the Americans had taken over 13,000 German soldiers prisoner.

Lloyd George learnt of the American success on his sickbed, immediately cabling Pershing that the news was 'better and infinitely more palatable than any physic.'[84] Ludendorff's only chance of avoiding defeat now rested with the well-fortified defences of the Hindenburg Line. However, with the reports of Pershing's victory only just beginning to reach the British public, and with Haig and Foch readying their armies for what could be the decisive encounter of the war, the last thing Sir William wanted to do was panic Britons about the health of their war leader.

In all, the Prime Minister's convalescence lasted ten days. For the first five days he was so ill that he did not have the strength to get out of bed and on Sir William's orders barriers were placed across Albert Square and the traffic diverted so that his patient would not be disturbed by the clanking of the trams – an indication perhaps that Lloyd George's symptoms included a painful headache, a common symptom of severe flu. The *Manchester Guardian* was not the only paper to downplay the seriousness of the Prime Minister's illness – the *Manchester Evening News* also blamed the weather, while *The Times* carried only the briefest mentions of Sir William's bulletins. On the sixth day of his illness Lloyd George was able to sit up in bed and receive visitors. But it was not until Thursday, 17 September, that Sir William finally allowed the Prime Minister out of the town hall for a short drive. However, there was no visit to Blackpool and when Lloyd George eventually left Manchester on Saturday, 21 September, in a private saloon car attached to the front of the London Express, he was wearing a respirator. On arrival at Euston, he was whisked directly to his official car, providing the *Manchester Guardian's* London correspondent, who was waiting at the platform's edge, with a brief glimpse of 'the familiar figure in the unfamiliar garb of a greatcoat, a large white muffler, and a soft hat turned down over the face.'[85]

With the help of the Prime Minister's aides and sympathetic newspaper editors, Sir William had done a good job of keeping the truth out of the public prints, but as the Secretary to the War Cabinet Sir Maurice Hankey noted in his diary soon after Lloyd George's return to London, the Prime Minister had been 'very seriously ill.' Lloyd George's valet, Newnham, also confided that it had been 'touch and go.' And writing to his wife Margaret from a colleague's country retreat after stopping briefly at Downing Street, Lloyd George confessed: 'I am crawling upward but have not yet recovered strength.' Even at the end of the month Lloyd George was still cancelling appointments, and when on 4 October he travelled to Paris for meetings on the war with other European leaders Sir William went with him. 'I am off by the 8 train from Charing X,' Lloyd George wrote to his wife early that morning. 'Sir William insists on accompanying me. My temperature is still very low & my pulse too feeble.'[86]

***

Lloyd George's flu is a mystery. It could have been a variation of the same strain that had been circulating in Manchester earlier in the year, or it could have been an entirely new strain, one that perhaps he had caught from a foreign delegation and carried with him by train to Manchester (in 1889–90 the first people to contract the Russian flu in Britain had been members of the diplomatic service, followed by prominent Lords and MPs, suggesting a key risk factor was contact with foreign embassies and consulates). Whatever the case, it appears that reports of the virus having abandoned Britain in August may have premature. Perhaps it had merely been biding its time, stealthily acquiring new mutations before bursting on the country with fresh lethality.

Just as no one can be certain where the spring wave came from, so no one can be sure where the second wave of the pandemic started. On 12 August 200 cases were reported on a Norwegian steamer at New York harbour, and four days later US public health officials ordered quarantine stations to inspect all vessels arriving from Europe following reports of 'so-called Spanish influenza' out of Valencia. At around the same time, the *HMS Mantua* arrived at Freetown, Sierra Leone, en route from southern England, with 200 sick sailors on board. Within a week 600 employees of the Sierra Leone Coaling Company had come down with the disease. However, as only a few died from pneumonia, and because the disease was generally considered mild, it was probably not a strain of the virus that had undergone critical step changes.[87]

Once again, the first unequivocal description of the emergence of a new, highly virulent strain of flu comes from a US Army camp. Towards the middle of August AEF surgeons reported that influenza cases were increasing in units in northern France. Then, in early September, the AEF reported several cases of influenza accompanied by 'severe pneumonia.' However, the critical outbreak occurred on 8 September when influenza erupted suddenly and without warning at Camp Devens, a US

Army facility 35 miles northwest of Boston. This new disease was incredibly virulent, and very different to the flu that had filled the infirmary at Camp Funston to overflowing the previous March.

'These men start with what appears to be an ordinary attack of LaGrippe [sic] or Influenza, and when brought to the Hosp. they very rapidly develop the most vicious type of Pneumonia that has ever been seen,' wrote Roy Grist, a member of the hospital's surgical team, in a letter to a colleague. 'Two hours after admission they have the Mahogany spots over the cheek bones, and a few hours later you can begin to see the Cyanosis extending from their ears and spreading all over the face, until it is hard to distinguish the coloured men from the white. It is only a matter of a few hours then until death comes, and it is simply a struggle for air until they suffocate. It is horrible.'[88]

The impact of this new 'grippe' was devastating. At Devens nearly 50,000 men were crowded into a facility designed for 40,000, with 5,000 living under canvas. The first man to fall ill was a soldier of Company B, 42nd infantry (the onset was so sudden he was initially misdiagnosed with meningitis). The following day, a dozen men in his unit fell sick, followed the next day by another thirty. Before too long daily hospital admissions were in treble digits and by the end of the month 14,000 men – or nearly a third of the camp population – had been hospitalized with influenza or pneumonia, and 757 were dead. As nurses ran out of beds, cots were laid along the three miles of hospital corridors to accommodate the overflow from the infirmary. In an era before antibiotics and vaccines, there was little Grist and his colleagues could do for the sick. The most effective treatment was aspirin and bed rest accompanied by attentive nursing. For patients with more severe symptoms doctors might prescribe quinine for relief of fever, morphine for pain, and digitalis, a drug that strengthens heart contractions and slows the heart rate. Other common 'remedies' included cinnamon, camphor, ammonia, eucalyptus and alcohol. But for those unlucky enough to develop pneumonia these were merely salves, the best advice being to get plenty of bed rest and to pray.

At the height of the outbreak Devens was averaging 100 deaths a day. 'One can stand it to see one, two, or twenty men die,' wrote Grist, 'but to see these poor devils dropping like flies sort of gets on your nerves.' An 'outrageous' number of doctors and nurses had also lost their lives, and, as the bodies piled up 'something fierce,' an extra barracks had to be vacated for use as a morgue. The pressure was very nearly unbearable:

> It would make any man sit up and take notice to walk down the long lines of dead soldiers all dressed and laid out in long double rows. We have no relief here, you get up in the morning at 5.30 and work steadily til about 9.30 P.M., sleep, then go at it again... The men here are all good fellows, but I get so damned sick of Pneumonia that when I go to eat I want to find some fellow who will not "Talk Shop"... We eat it, live it, sleep it, and dream it, to say nothing of breathing it 16 hours a day.[89]

Grist reckoned that the long lines of dead soldiers laid out in double rows 'beats any sight they ever had in France after a battle,' and certainly experienced pathologists, who were no strangers to gruesome battle wounds, flinched at the sight of victims' livid lungs frothing with bloody fluids. The disease struck pathologists as unique. As they moved a body to do an autopsy, 'a foamy, blood-stained liquid' oozed from the nose and mouth of the victim. On opening up the chest cavity, the lungs appeared blue and swollen 'with wet, foamy, shapeless surfaces,' and when they cut into the lungs they found 'intense congestion and haemorrhage.' For some medics the sight of the diseased lung tissue was 'too much.' It was as if the men had been visited by 'some new kind of infection or plague.'[90]

For patients, the symptoms were equally terrifying. 'You just felt as if your head was going to fall off,' reported an Army nurse who fell ill as she was about to sail for Europe on an AEF transport in late September. She spent the next ten days – almost the entire Atlantic crossing – in the sick bay. By now, so many nurses were stricken with flu that there was only one nurse to care for about 20 women. The nurse's most vivid recollection, however, was the odour. 'I never smelt anything like it before or since. It was awful, because there was poison in this virus.'[91]

In the febrile and highly charged political conditions that marked America's entry into the war it was not long before US newspapers were speculating that influenza was a new form of germ warfare put ashore by German U-boats. In fact, the evidence suggests that the virus travelled in the opposite direction and, as in April, it was American soldiers, disembarking at Brest, Portsmouth, Southampton, and Liverpool, who introduced the disease to France and England.

The American transports were the perfect vessel for the virus – floating incubators whose confined decks and overcrowded bunks revealed the flu's true ferocity. Conditions were worst of all on the *Leviathan*, a massive transport which left New York harbour for Brest on 29 September with 9,000 troops and 2,000 crew on board. The following morning every bunk in the sick bay was taken while other men lay prostrate in their berths. Within three days 700 men were sick and the ship had suffered its first fatality. Medics were run off their feet. As more and more men succumbed to the virus, doctors were forced to evict healthy men from their bunks, condemning them to less well-ventilated parts of the ship.

By 1 October more than 2,000 had succumbed and the scenes below deck resembled the aftermath of a pitched battle. According to the official navy report, 'pools of blood from severe nasal haemorrhages were scattered throughout the compartments, and the attendants were powerless to escape tracking through the mess, because of the narrow passages between the bunks.' There was little doctors could do except try to prevent patients from becoming chilled and keep them supplied with water and fresh fruit. But as the casualties mounted, even such minimal care proved beyond the *Leviathan's* capacity and as the decks became slick with blood and other bodily fluids 'the groans and cries of the terrified added to the confusion of the applicants clamouring for treatment.'[92]

By the time the *Leviathan* arrived at Brest on 8 October, some 2,000 men were ill with flu or pneumonia and more than 80 had died, the majority of the bodies having been disposed of at sea. By now everyone was desperate to get off the plague ship but 280 men were too sick to be landed and 14 more died on board that day. The remainder – some 1,000 men – were carried from the ship on stretchers in a convoy that extended for four miles from the wharf to the base camp. There they received further treatment. But as at Devens there was little doctors could do and by the time the illness had run its course the American casualties totalled several hundred.

As had occurred in the spring, next the virus spread rapidly from the AEF to the BEF and the French Army before crossing no-man's land to infect the German lines. According to Colonel Jefferson Kean, the deputy chief surgeon of the AEF, the disease's advance was frightening. 'Influenza increasing and spreading rapidly,' reads his diary entry for 28 September. Then, on 6 October: 'Influenza and pneumonia have increased by thousands of cases...the situation is very serious.' Four days later the situation had deteriorated still further with the AEF recording 1,451 deaths from flu. The majority, noted Kean, were the result of the associated pneumonias. Such infections killed nearly one man in two.[93] In all, the AEF would lose nearly 35,000 men to flu between September and the end of the pandemic the following April – almost as many men, in fact, as it would lose in the fighting.

**

Ironically, in spite of the alarming morbidity from flu in Britain in spring and early summer, the first nine months of 1918 had been a remarkably healthy time on the Home Front. Yes, food and fuel were in short supply, and yes, doctors had been alarmed at the poor physical condition of men reporting for national service – a finding that prompted Lloyd George to tell his audience at the Manchester Hippodrome that 'you cannot maintain an A1 empire with a C3 population' – but on most measures the nation's health had shown steady improvements since the turn of the century. The autumn wave of influenza, however, would change all that. In London alone, it would account for nearly 12,000 deaths between September and December, more if you take into account the 'excess' mortality during the same period from pneumonia and other respiratory diseases. The result was that 1918 would be the first year since records began that Britain's death rate exceeded its birth rate.

September had opened 'so wet and cold and windy,' according to the farming correspondent of one newspaper, 'that anyone might well think we had reached the fringe of winter.'[94] By October several rivers had spilled their banks and many low-lying fields lay flooded, delaying the corn harvest and lowering the yields of potatoes, turnips and other important sources of carbohydrate. As Britons hunkered down for a hungry and possibly disease-ridden winter, the nation's anxiety was reflected in a proliferation of advertisements for meat substitutes, disinfectants and tonics.

'Oxo fortifies the system against influenza action,' read an ad for the well-known beef-supplement in the *Tatler*. 'It will be apparent that a strong, healthy person will escape contagion when the ill-nourished one will fall a victim, consequently, one's aim must be the maintenance of strength.'

'Guard against influenza,' urged an ad for Jeyes fluid in the *Spectator*. '...spray the atmosphere of the office, factory, home [and] cinema... disinfect all lavatories, sinks and drains.'

One of the most popular ads of all was for an elixir called 'Veno's Lightning Cough Cure,' which a Lance Corporal in the 4th Essex Regiment swore had proved a 'wonderful' treatment for coughs and catarrhs:

> I was in hospital and lying opposite to me was a sergeant who had been badly gassed. It was awful to hear him coughing night and day...Knowing Veno's I told him of it. He got some, and from the first dose all the fellows in the ward noticed a decrease in his coughing.[95]

According to the corporal, in many Army camps 'heaps of fellows' were suffering from Spanish influenza. Yet he and a comrade had found 'instantaneous relief in Veno's.'

Interestingly, this obsession with influenza was not reflected in the editorial pages. On the contrary, most newspapers were reluctant to report that influenza had returned. The first mention came in the *Glasgow Herald*. On 3 October, A.K. Chalmers, Glasgow's medical officer of health, announced that there had been 66 deaths from influenza and 65 from pneumonia in the city. In actual fact, Chalmers had noticed the first increase in cases in the week ending 28 September, which suggests the virus had arrived on Clydeside about the same time as the American troopship *Olympia* was pulling into Southampton.

By 15 October Chalmers' figures made even more disturbing reading: 450 deaths from influenza and pneumonia, or 38 per 1,000 of population, the highest mortality rate recorded in Glasgow in 20 years.

It was a truly staggering statistic but as in many other British cities that autumn the authorities' response was phlegmatic. While Catholic schools in Glasgow bolted their doors, fearing that the crowded classrooms would spread the contagion from pupil to pupil, Chalmers argued there was little point in state schools following suit unless other places where people gathered – such as trams and railway cars – were also placed out of bounds. 'To select any particular service for drastic treatment while leaving the others untouched would mean tinkering with a problem which would not be worth the price,' explained the *Glasgow Herald* echoing Chalmers' advice. '[Such] drastic action... would recoil upon the community in a way which would make the cure probably worse than the disease.'[96]

If it had been the 'three-day fever' of old, there might have been some logic to Chalmers argument. But this was not the same disease that had caused children in Manchester to merely 'drop at their desks' in the summer. As many doctors noted the new strain was far more lethal. In London, for instance, it killed an average of 95 children a week versus an average of 11 a week in the summer. A popular playground skipping rhyme caught the ease with which this 'new disease' was transmitted:

I had a little bird
Its name was Enza
I opened the window
And in-flu-enza

As influenza spread from Glasgow to Aberdeen, and from Liverpool to London's Mile End, it was not long before playgrounds throughout the country echoed to the chant. By now influenza was everywhere. Carried in droplets expelled by sneezing and coughing, it was as ubiquitous as the air and as hard to avoid as breathing itself. In Abertillery, in Monmouthshire, the local medical officer of health reported that he knew of over two thousand cases, adding that 'the teachers think every other child has it.' In Grangetown School in Cardiff there were 1,000 cases and 14 out of 37 teachers were stricken. The authorities at Cheltenham College were so alarmed that instead of excluding pupils, they locked them in, forbidding them from approaching the town. In London, however, William Hamer, the Chief Medical Officer of the London County Council, deemed school closures impractical. 'We do not close schools in London as a matter of routine on account of influenza,' he rather pompously informed *The Times* in late October. Instead, Hamer issued instructions as to the 'precautionary steps' teachers could take in the event that flu symptoms became more widely prevalent.

The result was that while in other parts of the country schools bolted their doors, in the capital children were allowed to congregate in confined classrooms, thereby amplifying infections. Among the first to fall ill were schoolchildren in Lewisham, Lambeth and Brixton. Most likely this was because the south London boroughs were close to Waterloo and Charing Cross stations, the first alighting points for troops returning from France via Channel ports. However, within a week the flu had also spread to schools north of the river and there were soon reports of whole families being stricken. In Kingston, in southwest London, it was reported that two sisters had died on the same day and been buried in the same grave. Next came news of a five-year-old boy named Hatton who had collapsed and died in the street outside his family home in Edmonton. The source of the disease may have been

Edmonton Military Hospital where two nurses expired on the same day having reportedly caught the disease from patients in their care.[97]

In Islington and St Pancras medical officers printed handbills and placarded the streets with posters advising people to avoid crowds and keep bedrooms well ventilated and free of germs. In Poplar, where the epidemic was said to be 'very bad indeed,' the council distributed a powerful 'electrolytic' disinfectant with instructions to gargle and rinse the throat with it daily. Elsewhere, householders were told how to prepare their own gargles using permanganate of potassium and table salt. Most Londoners, however, refused to be fobbed off with such salves and before long doctors' surgeries and chemists were besieged by people demanding quinine.

The queues were particularly bad in Ilford. While the London County Council prevaricated over school closures, Ilford decided to bolt the gates of the borough's elementary and secondary schools, leaving 11,000 schoolchildren to kick their heels in the street. Inevitably, many flocked to cinemas, alarming moralists who considered the films shown in such venues a corrupting influence, and soon there were calls for the cinemas to be closed, or at the very least to be aired between performances. Scenting a marketing opportunity, some cinemas, such as Dally's in Leicester Square, decided not to wait for an official instruction and took out adverts boasting of the 'best ventilated theatre in London without draughts.'

On 22 October the *Daily Mirror* declared that flu was tightening its grip on the country with as many as 1,000 patients clamouring for treatment at some north London surgeries and many doctors also down with the disease. The following day the paper's tone was even more alarmist. 'The virulent disease is seizing victims wholesale and the doctors in London are unable to cope,' it declared. 'People are well one hour and down the next. Doctors and clients queues are the order of the day.'[98]

Even doctors hardened by hospital service were shocked by the severity of the symptoms. '[Cyanosis] developed in less than half of the pulmonary cases, but once it became definite the prognosis was so bad that I should say out of every 100 'blue' cases 95 died,' reported Dr Herbert French, the physician to Her Majesty's Household and Guy's Hospital. Even the mildest case had to be regarded as potentially grave.

> A patient might have been ill a day or two with mild influenza and seem to be progressing well [but] in an hour or two the whole picture might change, and twenty-four hours later the patient might be dead.[99]

Some victims – 'big strong men, heliotrope blue and breathing 50 to the minute' – would be fully conscious and clear-headed up to within half an hour of death, 'often not realising in the least how dire their condition was.' Others would slip into a delirium of the 'noisy, trembling, shouting-out type.' But the worst case by far, French wrote, was the type that became 'totally unconscious hours or even days before the end, restless in his coma, with head thrown back, mouth half open, a ghastly sallow pallor of the cyanosed face, purple lips and ears.' It was, he concluded 'a dreadful sight.'

In London, where nearly 61 people had collapsed in the streets in the space of 48 hours, by now these dreadful sights were commonplace. By the final week of October some 1,400 Metropolitan policemen were ill and in many parts of the capital there were so many firemen off sick that there were insufficient hands to man the pumps. The London Ambulance Service was similarly short-staffed and when some 1,000 telephone operators also fell ill the Postmaster-General

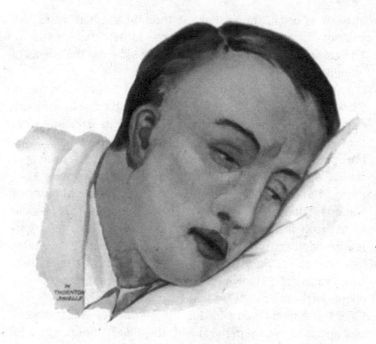

Figure 4.1 A patient with advanced cyanosis
Credit: Crown Copyright

issued an appeal asking subscribers to limit their calls to the exchange.

However, it was the personal tragedies that struck the deepest chord, like the case of two young would-be brides who worked in St Paul's Churchyard and had been planning a double wedding when they both contracted influenza and died.[100] Or the bizarre story of William Jones, a clerk for Nestle's Milk Company in Eastcheap, who instead of travelling to his desk in the City had caught a train from Paddington to Weymouth and thrown himself into the sea at Chesil Beach. At the inquest, his widow, who assumed her husband had been at work as usual, stated that she thought she had heard him call out her name and at once feared something was wrong. The coroner's verdict was 'suicide while suffering the effects of influenza.'[101]

As the fatalities increased, many funeral parlours ran out of wood for coffins and local authorities issued appeals for extra hands to bury the dead. Hackney solved its grave-digging shortage by co-opting employees from the town hall. Other boroughs used park gardeners. But in Bethnal Green, where bodies lay unattended in crowded dwellings for two weeks or more, the council decided to write direct to the Ministry of National Service. By now it was being reported that 'people were dying like sheep' and there were calls for the ministry to release doctors from the Army to help with the crisis.[102] Whitehall officials ignored the demands, preferring to refer such issues to the LGB. Only in Sheffield, it seems, did the ministry relent and agree to release soldiers to help with burials.

For those who could afford it, the best advice was to stay in bed and keep warm, perhaps with the aid of a hot poultice to prevent the lungs from becoming congested and a nursemaid to attend to the fever. But most working class families could not afford to hire nurses, even assuming they were available, nor could those employed in munitions factories or other essential industries easily take time off work. Furthermore, as news of the gruesome symptoms spread many caregivers were reluctant to approach patients too closely.

'If any of yr. household get the "flue," isolate the culprit & pass the food through the door!' Holcombe Ingleby MP wrote to his son from Norfolk in October. 'It is rather too deadly an edition of the scourge to treat it anything but seriously.'[103]

Curiously, few of the journalists, novelists and poets who lived through the pandemic and experienced the depredations of 'Spanish influenza' first hand thought to describe what it was like to suffer with the disease and certainly not in the sort of detail that allows us to envisage the horror of it now. Even as conscientious a diarist as Septimus Bennett, the brother of the social realist novelist Arnold Bennett, who as the manager of a Sheffield munitions factory kept a day-to-day record of his wartime experiences, struggled to put his impressions on paper.

'I had the most awful visions that I don't like to think about again, much less record,' reads his diary entry for 19 October. 'I have not been able to record anything or do anything worth doing in my spare time all week. Once or twice I made an attempt & failed absolutely.'[104]

There are good reasons for this. Flu is an extremely debilitating disease inducing a lethargy that can persist for days and sometimes weeks after an attack. In more extreme cases, flu can also provoke depression, psychosis and serious nervous complications. The result is that a person prostrate with flu is unlikely to have the energy or inclination to put pen to paper. Nor, once an attack has passed, do they usually wish to recall the experience for posterity.

Historians have also suggested other reasons for the flu's absence from public memory, including the callousness and indifference of a country hardened by four-and-half years of continuous fighting; the fact that most deaths from influenza occurred behind closed doors in the privacy of people's homes and were thus 'invisible' to public view; or Britons' legendary stoicism and the assumption that, however frightening the cyanosis and pneumonic complications, it was, in the end, 'only influenza.' But perhaps the most important reason for the absence of public memorializing was the sheer scale of the morbidity and the immense mortality associated with the pandemic. 'So vast was the catastrophe and so ubiquitous its prevalence that our minds, surfeited with the horrors of war, refused to realize,' wrote *The Times* in 1921. '…It came and went, a hurricane across the green fields of life, sweeping away our youth in hundreds of thousands and leaving behind it a toll of sickness and infirmity which will not be reckoned in this generation.'[105]

Such explanations are persuasive – up to a point. After nearly five years of unprecedented carnage during which the German 'sausage machine,' to recall Robert Graves's vivid phrase, had consumed the flower of British youth, the Great Flu must have seemed like a ghastly cosmic joke – God's last laugh, as it were, at mankind's expense. *The Times* was right – the influenza was everywhere and nowhere, ubiquitous yet invisible. This was not only a British phenomenon. In 1976 the American historian Alfred Crosby also remarked on the curious absence of references to influenza in the writings of American novelists and other prominent people who had survived the Great Flu, noting: 'One searches for explanations for the odd fact that Americans took little notice of the pandemic, and then quickly forgot whatever they did notice.[106] Crosby thought the main reason was timing: the vast majority of influenza deaths occurred in late October and early November, meaning they were overshadowed by the celebrations that marked the Armistice. And by the time the flu resumed its killing the following February, the US, like Britain, was focused on peace and reconstruction. 'Societies keep very poor records of why they do not think something is important,' he concluded.[107]

For those like Richard Collier who cared to look more deeply, however, there were plenty of people who were willing to share their experiences, and in 1972, when he began researching his book *The Plague of the Spanish Lady*, many of them were still alive. Their correspondence amounts to a secret history of the pandemic, a vivid picture of the impact of the disease on ordinary British households and a medical profession stretched to the limit. The letters also make it clear that for those who lost relatives, friends and colleagues the pandemic was an experience which seered itself in memory.

'In those days doctors wore top hats and travelled on foot, and these men, worn out with visiting patients, were to be seen tottering around,' recalled Bertram Copping, a newsagent's son from Islington, north London. 'Our doctor came, looked at my father, who one Saturday night complained of pains, and said he would call back in a day or two. When he did finally arrive, a week later, after repeated efforts on our part to reach him in the meantime, my father was dead.'[108]

Another correspondent, a GP's son from Fleetwood in Lancashire, painted an even more dramatic picture, recalling: 'So many were ill that only the worst could be visited. People collapsed in their homes, in the streets and at work. Many never regained consciousness. All treatment was futile.'[109]

Among the piles of correspondence Collier received, there were also poignant accounts of the disease's depredations. One came from a Mr Frewer in Carlisle. In 1918 Frewer was a 19-year-old officer in the London Scottish Regiment at Wisbech. The morning before he was due to rejoin his regiment he awoke to a temperature of 41° C. His mother immediately put him to bed, refusing to let the Army remove him to a military hospital.

'I was very ill for some weeks and eventually became certain that I could not possibly survive for the doctor had tried all known drugs (and I was becoming more and more depressed),' Frewer wrote. 'As a last resource he asked my mother whether I drank whiskey as, if I was not accustomed to it, it might have a shock effect on my body and cause the turning point. That night I had a half tumbler of whiskey and in a very few minutes was bleeding black blood from my mouth and nostrils…From that time I made slow but steady progress.'[110]

The expectoration of blood was not the only alarming symptom. Correspondents also describe the sudden loss of teeth and fingernails, of hair turning white overnight, and feverish dreams.

'I can remember having a very high temperature and we were all delirious, having terrible nightmares,' wrote Dorothy Jack from Kincairdshire, Scotland. 'There was too the awful feeling of complete prostration and I cannot find a better word to describe the feeling of misery and utter helplessness. Although I was aged 10 at the time this feeling was something I could never forget.'[111]

That phrase – 'never forget' – crops up time and again in the Collier correspondence. Writing from Coventry in 1973, Ethel Robson recalled how at the age of nine she was suddenly thrust into the role of sole caregiver for her family when her eight brothers and sisters, ranging in ages from 10 months to 15 years, contracted flu together with her mother. For some reason, Robson writes, 'I was the only one out of all the family that didn't have the virus.' Although a doctor visited twice a day, no one else was allowed into the house, 'therefore I was doing my best to help the others.' Thankfully, most of Robson's siblings recovered, but on 3 November her seven-year-old sister died, followed on 5 November by Robson's mother. The double funeral held on 11 November – Armistice Day – 'caused quite a sensation.'

> I can remember very well when the cortège was on its way to the church. Bells, hooters and all sounds of celebrations. It was raining but how silent people stood who realized it was our funeral. It really was a terrible time not knowing who we were going to lose next.

Afterwards, the doctor confessed to Robson that her brothers and sisters had been so ill that he had been fully expecting to sign five death certificates. 'I would like to tell you I am 64 years of age now but that period of my life I will never forget,' Robson informed Collier.

But amidst the loss and hardship there were also equally unforgettable acts of kindness. Mr Copping described how when his mother succumbed to influenza in addition to his father, 'relays of strange women came in, simply to sit by my father's bedside: it was just another case where the women spontaneously got themselves organized in the face of a common enemy.'

Doris Scott described how when she, her mother and her brothers and sisters came down with flu at their home in Stepney, East London, neighbours left bread for them on the widowledge. Others, like Edith Dilks, who contracted 'acute pneumonia' in the last week of October 1918, was fortunate in that a local Canon and landowner arranged for a woman to nurse her.

> She was a capable well-trained middle-aged woman and my husband says it was she who saved my life. Just remember there were no drugs...I was constantly poulticed and she made a cotton waistcoat...Throughout the crisis she was near me and she told me later, I clutched her waist with both hands and took all her vitality. She said she had a job to get a chair.[112]

By early November influenza had spread to virtually every town and village in Britain. Maurice Jago, a doctor's son from Land's End who had very nearly died of typhoid fever in 1914, recalled how during the seven long weeks the epidemic raged in Cornwall he rode from village to village with his father in a horse-drawn gig. 'St Buryan [the village next to Land's End] was primitive,' writes Jago.

> We had no running water, gas or electricity. All drinking water carried from central village source by hand. Oil lamps, privy

midden down the garden and at night the village lit by the parish lantern (moon)!... nearly all the patients were prostrate at home in bed, debilitated from war-time privation... I recall towards the end of that time father saying to me, 'My boy, in the last seven weeks I've been making money and if this is what "making money" [is] I hope to god I never make any more!!'[113]

The situation was little better in large inland towns. In Sheffield, for instance, where deaths were running at around 200 a week, doctors were so overwhelmed by new cases that the authorities took out ads appealing for volunteer nurses. Liverpool, Leicester, Portsmouth, Hull, Cardiff and Leeds were also badly affected and in Dublin, where 255 bodies were awaiting burial at local cemeteries, the authorities decided to disinfect the streets with Jeyes Fluid.

Meanwhile in London, it was reported that the flu's victims included Kingsley Doyle, the son of the novelist Arthur Conan Doyle. Kingsley had been wounded in the neck on the first day of the Battle of the Somme and had been making steady progress when he contracted influenza. He died of pneumonia in St Thomas's Hospital, London, on 28 October, prompting his father to turn to spiritualism for solace.

Kingsley's death seemed to underline the democratic nature of the disease. It didn't matter how rich your father was or how many servants or nurses you could afford, if the influenza 'germ' was going to get you then it was going to get you. 'Usually the advice as to how to keep from any plague is, "Have a large fixed income," and, in accordance with that maxim, one was told, before the war, to go to Cannes for the winter, to eat plenty of good food, to keep away from other people, and so forth,' commented the *Daily Mirror*. 'Now one is told to eat plenty of good food, not to work too much, and to keep out of crowds.'[114]

Such accounts bring home the suffering and tragedies behind the newspaper headlines. But for all that London and Britain's other great cities and towns were in the grip of influenza that autumn, it was a peculiar kind of plague, one that, for those who escaped its worst depredations, barely registered in the greater scheme of things. As the *Daily News* noted in early October:

London is more crowded today than at any time in the memory of the present generation, and now that the war news is so magnificently encouraging it is a brighter London too... The hotels are full; there are queues at the best and most popular restaurants; the shops are reaping a rich harvest; and the theatres and music halls are enjoying a boom.[115]

One can only conclude that this devil-may-care attitude was a product of the end of the war and the ubiquity of influenza, coupled with the fact for the vast majority of those infected flu was not an automatic death sentence. The result was that even when on 2 November it was reported that nearly 2,500 Londoners had died from flu the previous week and that the death rate was 55.5 per 1,000 – the highest since the 1849 cholera epidemic – there was little panic.

Perhaps this explains why the Local Government Board was able to get away with doing so little. As we have seen, in the summer of 1918 Sir Arthur Newsholme had shelved his plans to reduce overcrowding on trains and trams and to regulate public gatherings, reasoning that with the war raging such measures might harm the economy and that in any case there was little prospect of the War Office releasing doctors and nurses for civilian duties. The result was that it was not until 22 October that Newsholme issued a memorandum to local authorities with a covering letter suggesting that medical officers 'consider whether it would not be desirable... to prepare... some precautionary instructions.' These amounted to little more than a series of rules for patients, the first of which was to stay in bed for three or four days if they developed a severe cold or fever, thus reducing the risk of the disease being spread to work places. This was followed by a series of hygiene tips, such as avoiding sneezing and coughing in public and using a handkerchief to 'intercept drops of mucus' – the hand-kerchief being 'boiled, or burnt if of paper.' There was no need to disinfect every room in a house but it was advisable to clean patients' bedding and clothing. Because of the danger of relapse and the added complication of pneumonia later on in attacks, patients should also take care to keep warm and remain in bed until they were sure the fever had passed. Finally, they should consider gargling and irrigating the nasal passages with a 'weak' saline solution.[116]

The overall message, however, was that the medical profession was ignorant of the causes of influenza and powerless to prevent the epidemic. Furthermore, there was little point in making influenza notifiable, argued Newsholme, as patients frequently did not realize they had the disease for several days, and by the time they did it was too late for preventive measures. Nor was it incumbent on the LGB to lay on extra nursing cover. That should be provided by local sanitary authorities. In short, Newsholme concluded, influenza was a disease that could only be defeated by 'the active cooperation of each member of the community,' hence the stress placed in the LGB's memo on avoiding overcrowded dwellings and unventilated assembly rooms and discouraging 'indiscriminate expectoration' in public.

Warning that 'dirtiness' favoured infection, as did 'prolonged mental strain or over-fatigue, and still more alcoholism,' the memo made only a brief mention of the vaccine being developed by bacteriologists from mixed cultures of Pfeiffer's bacillus and the pneumococcus and streptoloccus found in victims' sputum and lungs. Trials of the vaccine had produced 'useful results,' but the public should not rely on it for widespread protection. 'This vaccine is being prepared in limited quantities, but is not as yet procurable for the present,' the memo concluded.

Newsholme's action did little to assuage his critics. The shortage of medical men was 'scandalous,' a doctor told an inquest in London. 'It was time more doctors were sent home from the front.'[117] Even *The Times*, which could normally be relied upon to toe the establishment line, criticized Newsholme's memo. 'It would have been better to lock the stable door before the escape of the horse,' the paper commented sarcastically. 'If this advice is likely to have any good effect, its chances of achieving its purpose would have been enhanced had it been published at the beginning instead of in the middle of the outbreak.'[118]

When it came to the LGB's defence that it could not have foreseen the recrudescence of influenza in the autumn *The Times's* verdict was even more coruscating. Pointing out that the LGB only had to recall the experience of the Russian flu to realize that a summer epidemic was likely to have been followed by a winter one of greater severity, it argued that the LGB had 'ample grounds for anxiety...even two months ago.' The paper concluded that the need for a Ministry of Health to coordinate the response to public disease threats had 'never received a more forcible illustration.'

Behind the scenes, Fletcher was also growing impatient. The day after the LGB published its memo he wrote to Newsholme renewing his call for help in investigating the epidemic and the 'dangerous sequels' to the primary attacks. Fletcher had begun to suspect that Pfeiffer's bacillus was a red herring and that influenza might be due to a 'filter-passing virus.'[119] The problem was that all the best scientific brains had been pressed into military service leaving the task of collecting bacteriological samples to 'scattered men over-burdened with other work.' The irony, Fletcher argued, was that the Army itself was now paying the price. 'Men they took away from what a year or two ago they thought of as "academic research" might by this time have done work of the first practical importance to the Army as such today,' he explained, before adding, somewhat perceptively in view of what we now know about the pathogenesis of influenza: 'I am trying hard to get one of two men back from routine Army work. My instinct is that the clue to influenza (so-called) will not come from work on the bacterial flora already recognised.'[120]

Fletcher's sense of urgency was driven not only by an appreciation of bacteriology but by his own experience of pneumonia in 1916 when, as he told Newsholme, the disease had nearly killed him and he was treated by 'mediaeval methods.' With the help of researchers at America's Rockefeller Institute and New York City's public health department – who had been investigating the cases of bronchopneumonia at Camp Funston and other US Army bases since the initial outbreaks in the spring – Fletcher had commissioned a study at Haslar Military Hospital in Portsmouth and Liverpool medical school. The idea was to collect serum from British flu victims before it could become contaminated with other bacteria and compare it with serum taken from American patients. At the same time, British Army and Navy researchers were pushing ahead with the production of experimental vaccines using the same blood and serum. Although Fletcher had his doubts about the efficacy of these vaccines, he considered them the best option for the time being, hence his plea to Newsholme: 'Could not your board arrange for a large supply being made available for the civilian population and supplied freely at hospitals and other centres?'

Although some vaccines were eventually forwarded by the LGB to Manchester, there is no record of Newsholme's response. Perhaps, as in August when he overlooked Fletcher's appeal in the *BMJ* for bacteriologists to send results of their investigations, Newsholme had more pressing matters to attend to. Instead, as criticism of the LGB mounted, the board convened a conference at which Fletcher and other leading experts, including Leishman and George Newman, the then Chief Medical Officer of the Board of Education, were invited to attach their names to a manifesto attesting to the LGB's performance. All the representatives, including Fletcher, declined. As Fletcher confided to the influential Conservative MP and aristocrat Waldorf Astor soon after: 'The main object of the conference appeared to be to secure some evidence which could be published, that the board had been active in relation to the epidemic.'[121] Newman's verdict was even more damning, recording in his diary that the meeting had been a 'futile waste of time' and that Newsholme was 'weak, vacillating, incompetent, untrustworthy & vain.'[122]

***

While the LGB's dithering was eroding public confidence in the medical profession on the Home front, the news from northern France was getting steadily better. Following the rout of the Germans at St-Mihiel, Pershing had advanced on a 33-mile front and by 16 September the Americans had taken several thousand German troops prisoner. Next, following a pre-arranged agreement with Foch, Pershing regrouped his forces and switched his point of attack north to the

Meuse-Argonne area. The idea was that the American First Army would fight in the Argonne forest to the right of the French Fourth Army, while the British – supported by the Canadians and the Belgians – would launch simultaneous assaults on the Hindenburg Line at Ypres and Cambrai. Unfortunately, the Argonne forest was far easier for the Germans to defend in depth and having advanced two miles Pershing was forced to pause while he rotated fresh troops into the front line. Nevertheless, by the middle of October he had broken through the main German defences and by the end of October he had driven the Germans from the forest.

The Allied attack to the north was even more dramatic. Supported by powerful artillery barrages, the Canadians managed to get across the Canal du Nord forming a breach that allowed the British Fourth Army under Rawlinson to strike against the toughest part of the Hindenburg Line. The heroes of the hour were troops of the 46th (North Midland) Division who overran German trenches on the near side of the waterway before swimming across and storming the German defences on the far side. Within a fortnight the Hindenburg Line was in ruins and by 9 October the British had retaken Cambrai. The joint British-Belgian assault on the old Ypres battlefield enjoyed similar success and within days the Menin Road Ridge, Passchendaele Ridge and all the familiar landmarks of four years of fighting were back in Allied hands.

The three-pronged Allied offensive was the beginning of the end for the Germans, forcing them to abandon the whole of the Belgian coast and much of French Flanders and setting off a process of collapse inside the German establishment. On 28 September Ludendorff reportedly flew into a rage, storming round his office and blaming everyone except himself for the defeats. The following day, he and Hindenburg held a meeting with the Kaiser at which they advised him to seek an armistice. Their demands triggered the resignation of the German Chancellor and the appointment of Prince Max of Baden, a political moderate, in his place. The result was that by the middle of October the new German government was involved in very public negotiations with the US President Woodrow Wilson over his 'Fourteen Points' peace plan and it was clear that an armistice was only weeks away.

There was no euphoria on the streets of London, however. Quite the opposite. It was as if after nearly five years of continuous warfare Londoners could not quite bring themselves to believe in peace. As Caroline Playne commented in her diary on 13 October: 'People look brighter but the thing is not yet quite believed or accepted. All the consequences of life adapted to war conditions have become so settled [that] you cannot break through & fully realise that a way out of the war may be found.'

As the populace steeled themselves for the possibility of further bad news, the influenza epidemic reinforced their sense of pessimism and gloom. 'Influenza very bad in places,' commented Playne on 26 October. 'Depression on faces very marked in trains and trams. People very full of sad cases of death from influenza. A great sense of dread about everything.'

Playne was not the only person struck by the melancholy air that pervaded the capital that autumn. Foreign visitors also remarked how Londoners seemed more preoccupied with influenza than the prospects of peace, prompting the *Guardian's* London correspondent to comment, 'People are afraid, even at this half-past eleventh hour, to let their thoughts rest on peace.'[123]

For years people had lived in fear of German occupation. The influenza played into the same sense of existential dread. It was as if the flu was the focus for all people's pent-up fears, all the stresses and strains they had spent nearly five years repressing. Besides, by now the smell of death was everywhere.

'There were families dying in [their] hundreds,' recalled one survivor from North Hillingdon in Middlesex. 'There was a coffin factory a few yards away from where we then lived and it was agony to see piles of unpolished coffins every hour up and down the street. People you saw one day were dead the next.'[124]

On 30 October, as the pressure mounted on the authorities to do something, the Ministry of National Service finally announced it was releasing doctors from the Army. It was a crucial decision. By now the epidemic was at its height with some 4,500 people a week dying of influenza in England and Wales. Although the LGB had yet to issue specific instructions, in Sheffield and Portsmouth the military authorities had declared cinemas and music halls out of bounds to troops. *The Times* did its best to calm the nation's nerves, arguing that although the pneumonia was of an 'extremely toxic character' there was no evidence that the epidemic was getting any worse. 'It is important to realize this and to see things in perspective as a stout heart is a great safeguard these days.' In a deliberate appeal to people's patriotic instincts it continued:

Fear is certainly the mother of infection. To go about expecting influenza is to invite it. Such an attitude lowers one's natural resistance, just as it lowers one's natural resistance to external enemies. The alarmists and the defeatists are the allies of the epidemic.'[125]

However, even *The Times* had to admit that there had been no improvement in London, and as October turned to November its obituary pages made for grim reading with the names of the civilian dead far out-numbering those of soldiers killed or wounded in action. On 29 October, for instance, *The Times* listed the names of some 80 civilians on its obituary pages. Nearly a third gave 'influenza' or 'pneumonia following influenza' as the cause of death. Some of the victims, such as Henrietta Vaughan, 'only daughter of Major and Mrs Charles Vaughan,' who was aged just nine were very young. But the majority were men and women in the prime life.

'At midnight on the 25th–26th October at his residence... of pneumonia after a few days illness,' reads a typical entry for a 49-year-old man from the East End. 'Cortege will leave the house at 2.30pm tomorrow... Friends please accept this (the only) intimation.'

Reading these announcements today, one cannot help but be struck by their terseness and the complete absence of sentimentality. There are no heroic epitaphs, none of the poetry that accompanies the entries for the fallen of Ypres and Passchendaele. It is as if the deaths from influenza are an afterthought of war – tragic, ironic even, but ultimately not something worth dwelling on.

***

If the mortality from influenza worried Lloyd George such concerns did not find their way into his letters or diaries. Nor do we know what theories, if any, he harboured about how he may have contracted the disease. As we shall see, James Niven, Manchester's MOH, thought that the second wave of the disease may have been brought to Manchester by American serviceman disembarking from transports in late September and early October, but, if so, that does not explain how the Prime Minister contracted the disease in early September. Nor does it explain why Manchester was one of the last major cities in Britain to be affected by the second wave.

In theory, influenza should have returned to Manchester at the same time as it hit other British cities. Together with the neighbouring borough of Salford, Manchester boasted a population of nearly one million, making it the second most populous urban area in the country after London. Manchester also enjoyed rapid rail links to Glasgow and the Midlands, and thanks to the Mersey ship canal running from Salford to Liverpool, any virus arriving by sea had a direct conduit almost to the heart of the city.

But when in the third week of September influenza broke out in Glasgow, Manchester miraculously escaped. A month later, deaths in London were running at about 1,500 a week. By now, influenza was also raging in other British cities, and Liverpool had registered 215 deaths. But on 23 October, next to a story about a woman falling from the top of an overcrowded bus, the *Manchester Evening News* recorded just eight deaths from influenza. The article advised: 'There are unmistakable indications that the disease is with us again... but [the cases] are being closely watched and there is no question at the moment of closing any schools.' The following day the paper's tone was even more confident, talking of Manchester's 'comparative immunity' from influenza and claiming that there was 'every likelihood' that the outbreak would leave Manchester 'almost untouched.'[126]

Niven knew that such talk was tempting fate and, sure enough, towards the end of October the deaths from influenza gradually began to creep upward. Alarmed, Niven had more handbills and posters printed recalling his earlier advice in the spring. 'Influenza is again prevalent in Manchester,' he warned. 'It is a highly infectious and very fatal disease, frequently leading to inflammation of the lungs...The disease is generally marked by its sudden onset ...wherever possible the following precautions should be taken.'[127]

Niven then spelled out how people should avoid crowds and in the event that they fell sick should isolate themselves, preferably in a separate room ('in this way the spread of the disease will be delayed,' he explained). Once attacked, patients should stay in bed, keep warm and immediately call a doctor – 'any delay may result in dangerous complications.' Discharges from the nose and mouth should be also destroyed, if possible 'in a clean rag or paper, which should then be burnt.' Finally, in capital letters Niven spelled out how 'THOSE ATTACKED EVEN SLIGHTLY SHOULD ON NO ACCOUNT JOIN ASSEMBLAGES OF PEOPLE FOR AT LEAST A PERIOD OF TEN DAYS FROM THE COMMENCEMENT OF AN ATTACK, AS THEY MAY CONVEY THE DISEASE TO OTHERS.'[128]

It was sound advice but for all that Niven's appeals may have delayed the spread of the disease they could not halt it, and when on the morning of Tuesday, 11 November, the news came that Germany has signed the Armistice the inevitable happened. Unlike the day exactly two months before when Mancunians had cheered the Prime Minister's procession through Manchester in the rain, Armistice Day had broken fair with what the *Manchester Guardian* described as 'a yellow autumn sunshine…coining the wet streets into gold.'[129] Many people had stood vigil all night outside the newspaper's offices and when at 10.25am the Union flag was raised above the building confirming the peace, office workers all over the city spontaneously opened their windows in relief. By midday flags were flying from nearly every building in the city and by 1pm hundreds of female munitions workers began streaming towards Albert Square, singing patriotic songs as they marched. Their joy was infectious and they were soon joined by British, American and Belgian soldiers, as well as trolley guards who had abandoned their posts. By early afternoon Albert Square was pulsating. The munitions workers swarmed the entrance to the town hall,

Figure 4.2 James Niven
Credit: Wellcome Library, London

their clogs clattering merrily on the tessellated paving stones, while outside soldiers broke into foxtrots as office workers cheered them from the deck of a No. 38 omnibus that had become marooned in the crowds.

By early evening, with the tram system at a virtual standstill, the crowds began fanning out into the surrounding streets, linking hands across the roadway behind bugle bands and drummers. The celebrations made a 'wonderful' contrast with the recent dark days, wrote the *Manchester Guardian*, the crowds moving 'as if they had no care and no thought beyond the burning joy of the moment,' and come nightfall, with no way of getting home, people packed into musical halls and theatres where they entertained each other with further singing and flag-waving into the small hours.[130]

The next day, to everyone's surprise, the party continued. The Munitions Works had declared Wednesday a holiday and the mills quickly followed suit so that by midday the Manchester crowds had been further swelled by 'the incursion of innumerable hosts of people' from the outlying Lancashire mill towns.[131]

It was Niven's worst nightmare. By Wednesday afternoon the main thoroughfares of the city were so jam-packed with people that the trams were once again brought to a standstill. That day the *Manchester Evening News* reported that the death toll had risen to 149 and that because of the crowds it was likely that the next few days would also show a further increase. 'By leaving their homes it is quite probable that many millions of microbes have been passed from one to another,' the newspaper warned.[132] But no one was listening.

On 18 November the newspaper carried the headline 'Alarming Increase in Manchester Mortality.' But it was not until the final week of November that the fatalities shot off the chart and that, as Niven put it, 'it was realized that a real calamity had befallen the city.' In that week alone Niven recorded 383 deaths, more than double the number of the final week of the June outbreak. The suffering was terrible.

> Mothers and fathers were often stricken together. The children, themselves ill, could not receive attention, and for a time it seemed as if it would not be possible to get coffins for the dead, or grave-diggers to dig the graves... Bodies were left as long as a fortnight unburied, partly at home, partly at mortuaries, and partly at the premises of undertakers.[133]

One of the first casualties was a seven-year-old girl, Ada Darwin. In November 1918 Darwin shared a small two-up, two-down house with her mother, Jane, and her four brothers and sisters in Birch Street, not far from Maine Road in South Manchester where the Manchester City football stadium stands today. Darwin's father, Frederick Berry, had served in the Boer War and on the outbreak of World War I had volunteered for the RAMC, leaving Ada and her siblings – Austin, 2, Noel, 4, Norah, 12, Frederick, 9 – to subsist on his Army pay. It was a typical working class neighbourhood. 'I'd seen ragged barefoot children in areas near to Manchester City, places involved in the mill industry,' says Darwin. 'We ourselves were 'poor'. Our clothes mainly were made by grandma, mother and aunties.'[134]

Several of her aunts worked in service, in the big houses of the 'new rich' – the owners of the cotton mills and textile factories that lined the Manchester ship canal. Their contributions helped supplement the money Jane received from Frederick's meagre Army salary, ensuring that Darwin and her brothers and sisters had enough 'not to be actually hungry.' On Saturdays they received one penny each for the cinema and a ha'penny for sweets. Otherwise they spent the time skipping or playing with tops and hoops in the street, imagining that the 'pretend' shops lined with empty containers contained tropical fruits and other exotic foreign produce.

Armistice Day, says Darwin, dawned 'rather suddenly.' 'I remember how pleased mother was saying our dad would not be off away from home any more.' But rather than resigning his commission Frederick decided to stay on at Salford Military Hospital tending to the sick and wounded. It wasn't long after, Darwin says, that the 'joy of peace was overtaken by the increase of influenza.'

She was the first to fall ill. 'It was Sunday, November the 17, that I was put to bed. I remember this great big headache and telling my mother to stop my sister Nora chattering, it's making my head hurt.'

Darwin cannot be sure but thinks that it was her mother, who had been nursing a sick neighbour, who had brought the virus into the house. Darwin was put in her parent's bed. Soon afterward, she was joined by her mother and baby sister, Edith, followed by her younger brothers, Noel and Austin.

'Then I must have got over the worst because I can remember sitting in the big iron cot with Noel and giving him his medicine. It smelt just like TCP.' Her grandmother, who was a nurse, arrived soon after, having been alerted by Darwin's eldest brother, Fred, who was worried that his mother had left washing in the boiler and there was no one to take it out. 'My grandmother said she knew straightaway that my mother was going to die because there were black patches on her.' These were the characteristic marks of heliotrope cyanosis.

On the Monday the doctor arrived and on his advice Darwin and her brother Austin were sent to their aunt's house to be nursed away from the rest of the family, while Frederick and Norah were sent to their grandmother's. That was the Tuesday. Darwin's mother died the next day – Wednesday 20 November, followed on the 23rd by Noel. Five days later, on 25 November, Darwin learnt that her father, Frederick Berry, had also died. He had apparently caught the flu while nursing patients at Salford Military Hospital, dying in the early hours while calling for his wife.

Berry was buried with full military honours at Manchester Southern Cemetery on 29 November. Only the older Berry children were allowed to attend but Darwin, now aged 96 and living in Chester, can still recall watching his funeral cortege pass near her primary school.

> It's like a film in my head. There were the black horses with the plumes made from ostrich feathers, then the gun carriage with my dad's coffin covered with the union flag. My mother's coffin was in a big glass hearse with Noel's coffin under the driver's seat. My grandma told us my mother had gone to Jesus, but I said, 'Jesus has got plenty of people, I want my mummy.'

Niven had little doubt that the Armistice Day celebrations were to blame for the upsurge in flu cases and was determined not to give the virus further opportunities. In the second week of November the LGB had finally got round to issuing regulations requiring cinemas, many of which screened films continuously from lunch until late in the evening, to be cleared and ventilated for at least 30 minutes between performances. On 20 November Niven urged the Manchester Watch Committee to enforce the new regulations, arguing that in the course of a three-hour entertainment 'much disease' could be passed from one person to another and that the majority of cinema houses were 'vile.'

Niven's arguments were persuasive, but only up to a point. After hearing from cinema owners that the restrictions would force their clients to wait outside in the cold and rain, putting their health at possibly greater risk, the committee compromised by recommending just a 15-minute pause between performances for ventilation.

Niven also tried to get the committee to address the problem of the crowding on trams and trains, arguing that in a confined car or compartment a man ill with flu 'will pass his germs in far greater numbers to [other] passengers.' However, in this he was rather less successful. Nor, despite his handbills and his warning about the dangers of sputum, could he do much to stop people spitting in the streets.

By the end of November deaths were running at more than double the birth rate. As the bodies piled up, Manchester followed Bethnal Green's example and appealed to the National Service Board to release soldiers to help with the burials. Eventually the services of a detachment of the Labour Corps from the Western Command was secured. Even so, coffin-makers struggled to keep pace with demand and soon funeral parlours were calling on families to consider cremation instead.

On 9 December, beneath the headline 'At Last!,' the *Manchester Evening News* reported a drop in the death rate, with 243 fatal cases compared to 383 the week before. But, once again, the newspaper was overly optimistic and it was not until the first week of January that Niven finally declared the epidemic over.

In all, 1,715 Mancunians perished in the second wave of the pandemic, with women accounting for two thirds of the deaths. At the peak of the outbreak, in the final week of November, the death rate had reached 46 per 1,000, the highest level since the 1849 cholera. In London, where the death rate in some boroughs had been as high as 92 per 1,000, the toll had been even greater, with as many as 16,000 deaths according to some estimates.[135]

But just as worrying were the pattern of the attacks and the profile of the victims. In 1891, those who had been attacked during the first wave of Russian flu in the winter of 1890 had enjoyed immunity when the disease returned in the spring. But in the autumn of 1918 Niven remarked that those who'd been ill with influenza the previous spring appeared to enjoy no such protection. Like others, Niven also remarked how it was the healthiest and most vigorous members of the population who seemed to be the worst affected.

'When curves of frequency of deaths at different age periods are constructed it is found that they all present a striking peak at ages 25–34,' he wrote. 'This is totally unlike the behaviour of influenza in previous pandemics and requires special study.' Like other commentators, Niven thought the phenomenon might be connected to the war. 'It is possible that the aggregation of young adults may have had something to do with it...coupled with the sorting out of less vigorous men.'

He also could not help noticing that the death rate from influenza in South Manchester had been exceptionally high, certainly far higher than could have been predicted from the usual death rates for respiratory diseases. Shortly before the second wave had hit Manchester, Niven pointed out, large numbers of sick American serviceman had been taken to the Old Trafford cricket ground in southwest Manchester for treatment. Perhaps, he suggested, the two were connected:

Large numbers of American soldiers visited this country and France in September and October, and its known that some ships witnessed very painful scenes, the men suffering from influenza in a very severe form. It appears likely that this outbreak was introduced from America and that increase virulence was imported at the same time.[136]

5

third wave, January–May 1919

On 17 January 1919 the headline in the *Manchester Evening News* read 'Vanishing Influenza.' As the prospect of a new and more peaceful era dawned, the Spanish influenza appeared to have gone the way of all plagues of war. Confident the worst was over, Niven advised the education board it was safe to reopen the city's schools and by the end of the month playgrounds in South Manchester once again reverberated to the snap of skipping ropes and innocent children's rhymes. But the worst was not over. As in August, the surface proteins of the virus – the cloak-and-dagger-act of haemagglutinin spikes and mushroom-like neuraminidase – were merely drifting before re-emerging in altered form. In 1919, of course, no one had the faintest idea what a virus was. All they had was guesswork and educated hunches.

The notion of the existence of microbial forms too small to be detected by optical microscopes dated back to the late 1870s and the Irish physicist John Tyndall's experiments on 'floating matter in air.' Tyndall designed a special chamber containing unfiltered and filtered air, then introduced meat into the chamber. In the 'optically pure' or filtered air the meat retained its colour, but in the unfiltered air it turned putrid. Furthermore, when Tyndall shone a concentrated beam of light on the chamber he could see tiny particles in the diffracted light – particles too small to be seen through a conventional light microscope. From this Tyndall drew the inference that air must contain submicroscopic infective organisms – what other scientists had hitherto dismissed as merely 'potential' or 'hypothetical' germs.

With the invention of improved achromatic lenses and better culture staining techniques, by the late 1880s Louis Pasteur and Robert Koch had brought a series of hitherto hard-to-detect germs into view. These included not only such landmark bacteria as the bacilli of fowl cholera and tuberculosis, but the three most common bacteria found in the nose, throat and respiratory tract: namely, pneumococcus, staphylococcus and streptococcus. Another bacterium that was just big enough to be seen through a light microscope was Pfieffer's bacillus, *H. influenzae*. Like other common bacteria of the respiratory tract, *H. influenzae* could also be filtered from sputum and cultured on blood agar. But there were some infectious organisms that could not be cultured no matter how hard bacteriologists tried and because they were smaller than the wavelength of light they couldn't be seen through a microscope either.

One example was smallpox, another was rabies. In the 1880s Pasteur had tried to culture rabies repeatedly without success. Other scientists had attempted to do the same for smallpox, a disease against which Edward Jenner had pioneered a vaccine a century earlier. As with rabies the attempts failed for the good reason that both smallpox and rabies are transmitted by viruses not bacteria. Rather than being deterred by his failure to grow rabies *in vitro*, however, Pasteur simply injected the infected material directly into the brains of dogs by trepanation. The agent was then passed through rabbits by successive inoculation to increase its virulence before being removed from the rabbit's spinal cord to obtain what Pasteur called a 'fixed virus.' Having ascertained the maximum virulence, Pasteur then attenuated his viral culture before administering the vaccine to animals and, eventually, humans. Gradually, using vaccines of different strengths he hit on a level of the virulence that was both safe and which augmented resistance to the disease.

Pasteur's success opened the way for the development of vaccines for a host of bacterial and viral diseases and the elucidation of the existence of other viruses. Using fine linens or a hollow porcelain cylinder known as a Chamberland filter, bacteriologists repeatedly filtered infected material and then checked it for bacterial cultures. Only when they were sure the filtrate was free of bacteria did they reintroduce the 'filter-passing' virus into plants and animals. Employing this method, by the turn of the century scientists had elucidated the existence of both the tobacco mosaic virus and the virus of foot-and-mouth disease. And by the outbreak of the war scientists had developed vaccines against cattle plague, typhus and yellow fever.

However, while such vaccines were a key public health weapon and fuelled important insights into immunology, scientists did not use the term 'virus' in the sense that we use it today. In particular, bacteriologists had no concept of how viruses invade and take over the machinery of animal cells in order to replicate and make multiple copies of themselves. Instead, they tended to conceive of such filter-passers as specialized forms of bacterial 'poison' – ultramicrobes that had evolved from bacteria by increasing parasitism and that like larger bacteria multiplied by means of binary fission. The best they could hope for was a bacterial vaccine or a passive blood serum. But which bacteria should they seek to culture – *H. influenzae* or one of the several strains of pneumococci, streptococci and staphylococci commonly found in patients' nasal passages and lungs?

Questions about *H. influenzae's* role in the aetiology of influenza had dogged the bacillus every since Pfieffer's announcement in 1892 at the height of the Russian flu pandemic. Pfieffer had isolated the tiny Gram-negative bacillus from the sputum and lungs of sick patients and had subsequently succeeded in growing it in a pure state on an artificial medium, thus meeting Koch's second postulate used to test for the aetiology of disease. But when Pfeiffer came to inoculate monkeys with the bacillus none of them developed conclusive clinical symptoms of influenza, thus failing Koch's third test. Not only that but Pfieffer himself recognized that even during the Russian flu epidemic he had found the bacillus in only half the cases he had investigated.

Nevertheless, the close association of the bacillus with the lesions of the respiratory tract in human influenza cases and the frequency with which it was found in patients' sputum were highly suggestive of a connection and most bacteriologists decided to give Pfieffer the benefit of the doubt. Besides, by 1884 the Russian flu had run its course and there were fewer opportunities to culture the bacillus and test Pfieffer's claims. It was only with the emergence of a new pandemic strain in the spring of 1918 that widespread studies became possible again and doubts resurfaced. Time and again bacteriologists tried to culture *H. influenzae* from sputum or secretions of the lungs and respiratory tract, only to discover the bacillus was present in miniscule quantities or completely absent. Convinced their filters were at fault or they were using the wrong medium to grow the cultures, they tried again only to come up with similar results.

By the summer of 1918 the concerns about the aetiology of influenza had reached such a pitch that a special meeting was held at the Munich Medical Union. Summarizing the union's conclusions the *Lancet* wrote that 'Pfieffer's bacillus has been found but exceptionally' and that if any bacteria had a claim to be the cause of influenza it should be the far more common streptococci and pneumococci. The following October the

Lancet was in a somewhat more generous mood, writing that 'the question whether Pfieffer's bacillus is the specific cause of influenza is still an open one' and that it was unfortunate that researchers did not have better agglutination tests for the 'micro-organism.'[137] By now, bacteriologists were having rather more luck culturing the bacillus, though whether this was due to the use of better blood agar mediums or the interaction between it and other bacteria during the second wave of infections is unclear. For all that the bacillus was being found more frequently, however, the Royal College of Physicians had joined the doubters arguing that there was 'insufficient evidence' for Pfieffer's claim, though it was happy to allow that the bacillus played an important secondary role in fatal respiratory complications of influenza.[138] For all that the medical consensus was moving against Pfieffer, however, many scientists, including Britain's foremost pathologist Sir Bernard Spilsbury – known as the Sherlock Holmes of medical detection because of his celebrated use of forensic techniques in murder cases – continued to claim that the fault lay with scientists and their equipment. Indeed, even after conducting numerous post-mortems on influenza victims, during which he had succeeded in isolating H. influenzae just 40 per cent of the time, Spilsbury maintained that the absence of the bacillus was probably due to his failure to look closely enough.[139] Another doctor writing to the BMJ on Armistice Day acknowledged he had discovered the bacillus in some epidemics but not others. He then went on to claim he had developed a foolproof method of culturing the bacteria before concluding, somewhat unsatisfactorily, that when the bacillus was 'constantly' present it was 'either a cause or an effect' of the disease.[140]

Given the doubts over the role played by H. influenzae and other bacteria in influenza's aetiology one wonders why the authorities bothered to develop any vaccines – indeed, the doctor just cited thought the effort pointless, arguing that until the cause of influenza was known such vaccines 'would probably do more harm than good.' He was almost certainly correct. Even today, employing egg-based vaccine production technology, it takes a minimum of four to six months from the isolation of a new strain of flu to the manufacture of a prophylactic vaccine. But in 1918, scientists had no reliable means of isolating the virus from patients, much less of identifying key antibodies in the blood and producing mass quantities of vaccine. Even if they got lucky and succeeded in obtaining secretions of sufficient virulence to enable trials to be conducted in animals there was no guarantee that the resulting vaccine would confer protection against the prevalent strain of influenza. Finally, there was the question of the degree to which any vaccine could provide protection given the wide range of immunological responses observed in the general population.

Following the second wave in Manchester, for instance, James Niven had ordered officials to carry out a block survey of 1,000 households to ascertain how many of those attacked in the autumn had also been attacked in summer. The idea was to establish whether those who had survived earlier attacks enjoyed immunity against subsequent attacks. The results were equivocal. Some 500 households – or half those surveyed – had been invaded in either the summer or autumn and, of these, 102 had suffered in both epidemics. In all, 480 people had been attacked in the autumn, including 73 who had been attacked in the summer. In other words 37 per cent of those who should have been protected were not. In a survey of British soldiers hospitalized in northern France in October Major Greenwood reported similarly equivocal findings: out of 965 men hospitalized, 248 – or little over a quarter – had also suffered influenza in the summer. While Greenwood argued

Figure 5.1 Major Greenwood
Credit: Wellcome Library, London

that this showed the summer attacks had conferred a 'measure' of protection he acknowledged the protection was weak. Niven's verdict was even more pessimistic. The results of the Manchester block survey, he argued, 'could only be interpreted as showing that one epidemic outbreak gave no protection against the next.'[141]

For all the doubts about the utility of a vaccine, however, the Army was convinced it was worth a try. The key decision had come at a conference at the War Office on 14 October attended by leading military and civilian bacteriologists and pathologists and chaired by Sir William Leishman, the Director General of the Army Medical Services. Despite the continuing doubts over influenza's aetiology, Leishman claimed there was no question of *H. influenzae's* role in the production of symptoms and secondary complications. Similar arguments applied to the pneumococci and streptococci frequently found in the nasal passages and respiratory tract of influenza victims and which were thought to be the cause of severe secondary pulmonary complications. Leishman argued that these bacteria should now be added to the mix alongside Pfieffer's bacillus in the hope that one or other, or else all the strains of bacteria in combination, might confer a degree of protection against future attacks. In view of the continuation of the influenza epidemic into the winter, the development of such a polyvalent bacterial vaccine was of 'great and urgent importance,' he concluded.[142]

Leishman's main motivation was to produce a prototype vaccine that could be deployed by the Army in case of an extension of the war, and although there is evidence that Fletcher shared the doubts of other scientists about the utility of such a vaccine he decided to go along with the proposal in the hope of further elucidating the aetiology and pathology of the disease. Spurred by the illness of several colleagues,

Fletcher also realized that there was a need to obtain results 'quickly while the opportunity lasts' as the epidemic would soon be over and the chances to gather usable infective material would then be lost. Accordingly, he agreed to help Leishman coordinate the work of military and civilian bacteriologists in the hope that should the Army's vaccine research succeed it could also be rolled out to civilians. One of the sites selected for the trials was the Royal Naval College in Greenwich. The other was Abbeville, an Army laboratory 50 miles south of Etaples, at the mouth of the Somme.[143]

On the reappearance of influenza in the BEF in October, Major-General Sir John Rose Bradford, the consultant physician who had first drawn medics' attention to the outbreaks of purulent bronchitis at Etaples in the winter of 1917, had assembled a crack group of bacteriologists, pathologists and clinicians to investigate the aetiology of influenza and other diseases suspected of being due to filter-passing viruses, such as poliomyelitis and nephritis. The research team was headed by Colonel S.L. Cummins, adviser in pathology to the British Armies in France, but the actual experiments at Abbeville were conducted by Major Graeme Gibson and two other men, Major F.B. Bowman and Captain J.I. Connor.

By now similar trials were also underway in other countries. At Chelsea Naval Hospital in Boston, for instance, Lieutenant Commander Milton Rosenau, a future president of the Society of American Bacteriologists, had begun extensive experiments with 62 volunteers from a navy brig. Collecting sputum and blood from influenza patients he passed the drained-off fluid through a porcelain filter then tried to communicate the disease to the volunteers by injection, inhalation and by dripping the fluid directly into their nasal and throat passages. But Rosenau had been unable to transmit the disease. In Germany, scientists were conducting similar tests, spraying themselves with secreted material from the nose and throat of influenza patients that had been filtered so as to remove bacteria, but they too failed to contract influenza. Only in Paris were scientists reporting any success. In early October Charles Nicolle and his colleague Charles Lebailly took unfiltered bronchial secretions from patients with influenza and injected them into the eyes and nasal passages of macaque monkeys. After six days the animals developed fevers lasting three days. However, when they repeated the experiment on two human volunteers only one fell ill. A few weeks later another French researcher tried again, this time on himself. Taking 20 cubic centimetres of blood from four patients he mixed the blood products together and passed them through a Chamberland filter before injecting himself with 4 cubic centimetres of the serum. Four days later he suffered a severe attack of influenza.

Because of the severity of the influenza and the risks involved Gibson decided not to repeat the experiments on human volunteers. Instead, he decided to test the sera on rabbits, mice, baboons and rhesus monkeys, looking to the lesions in the animals' lungs to indicate that a positive result had been obtained.

The rabbits and mice were relatively easy to come by, the monkeys less so. With a world war raging primates couldn't simply be airlifted to France from the jungles of Africa. Instead, Cummins enlisted Fletcher's help in rounding up monkeys from government labs, zoos and private collections. The effort was complicated not only by the shortage of supply but by the monkeys' reluctance to cooperate, as was evidenced on 5 December when a monkey escaped its Home Office minders and went walkabout in Westminster. The next day, Fletcher wrote to Cummins to say that the pesky primate had been spotted at New Scotland Yard, 'presumably about to report himself to the police.' Unfortunately, the monkey evaded arrest prompting a policeman to pursue him as far as Whitehall where the animal was run over by a bus. Thinking the chase over and his quarry dead the policeman went to retrieve the corpse only for the monkey to suddenly spring to life and scramble up the façade of the Home Office. The primate's exploits delighted the assembled crowd but it was his last hurrah. 'He was found dead at the top of the Home Office that evening,' Fletcher informed Cummins, 'dead, but not dishonoured.'[144]

By now Gibson had infected several rabbits, mice and monkeys both with sputum and blood and was reporting that he had succeeded in producing lung lesions 'similar to those seen in human cases.' Fletcher thought the result 'very interesting and important' and urged him to see if he could repeat the experiment. 'I do hope you will go "all out" in this work, regardless of the cost, which, however big it will be, would be negligible in relation to what is at stake...' To confirm the hypothesis that the infectious agent was a virus, Fletcher suggest Gibson heat half the filtrate to 55° C for 30 minutes, and then inject one monkey with raw filtrate and the other with the heated virus, 'to see if the latter escapes when the other is positive.'[145]

On 17 December Gibson wrote to Fletcher that he was having difficulty reproducing the microscopic lesions in monkeys' lungs that he'd seen in earlier experiments and conjectured that the failure might be due to the fact that the influenza virus was decreasing in virulence. Nevertheless, he urged Fletcher to send him three dozen extra mice as he thought he'd got the 'filterability of [the] virus proved.'

By now, the British were dismantling their camps in northern France and funds were drying up, threatening not only Gibson's research but the parallel experiments being conducted by Bradford at No. 20 General Hospital at Etaples. The end of the

Figure 5.2 A mobile bacteriology laboratory. Designed by Baird and Tatlock (London) Ltd. It was presented to the Army Medical Services in France by HRH Princess Christian.
Credit: Wellcome Library, London

war and the decline in influenza cases was also threatening to foreshorten the Army's trials with an experimental passive immune serum. So far some 2,000 troops had been inoculated. Only 2 per cent had developed influenza but since a similar proportion of the control group, consisting of twice as many soldiers, also developed influenza the results were deemed inconclusive.

In Britain, Fletcher was facing a similar struggle to sustain interest. Although the MRC had an annual budget of £57,000 a year – about £4 million in today's money – the lion's share was earmarked for tuberculosis research and with influenza cases falling Fletcher was under pressure to focus on broader public health concerns. Following the meeting at the War Office he had recruited a team of researchers at British military and civilian hospitals and the labs of several American hospitals. Convinced that only a broad spectrum approach to a vaccine held out any hope of success Fletcher arranged for the Rockefeller Institute in the US to forward sera from American influenza patients containing several different strains of pneumococcus. He then asked British researchers to compare the sera with strains collected from British influenza patients to see if there was any affinity. As he had told Newsholme two weeks after the conference at the War Office: 'My instinct is that the clue to influenza (so-called) will not come from work on the bacterial flora already recognized.'[146]

But by January Fletcher too was finding it increasingly difficult to obtain serum. Then, in February, came the news that Gibson had paid the ultimate price. According to Cummins, Gibson had been working long hours in the lab at Abbeville, repeatedly filtering material to be sure it was free of bacteria and conducting passage experiments, when he was attacked by the disease in its 'severest form.' Given the difficulty American researchers had in passing the virus to volunteers it seems unlikely that Gibson had accidentally infected himself. Probably, he was tired and worn out and simply fell victim to an opportunistic infection. Whatever the case, Cummins wrote that Gibson's death was a 'grievous loss' to medical science as he had been on the point of completing his experiments into what he 'believed to be the "Filterable Virus".' Nevertheless, it was 'such an end as a soldier would have chosen,' showing, Cummins concluded, that 'dangers resolutely faced are no less glorious in the laboratory than in the trenches.'[147]

Happily, other researchers enjoyed better luck. At the Royal Naval College in Greenwich, where Surgeon-Captain Percy Bassett-Smith cultured a mixed vaccine containing several different strains of each of the three main bacteria, 1,000 boys at the Royal Hospital School were inoculated. None went on to develop influenza even thought the disease was very prevalent in the area. The vaccine enjoyed rather more mixed results when medical offices came to test it on volunteers in the Grand Fleet, with ten men developing mild symptoms of influenza on the first day. 'The preventive value of the vaccine against influenza bacilli is not high,' Bassett-Smith concluded, 'but it is probable that an increased immunity against strepto-cocci and pneumococci infections will be acquired, preventing the most severe complications and fatal results.'[148]

By now even Newsholme had become a reluctant convert, forwarding a phial of experimental vaccine to Niven in Manchester. Niven sent it on to Monsall Hospital where it was given to five nurses who were continually exposed to influenza patients. Just one nurse developed influenza 15 days later; the other four escaped infection.

At the summit meeting on influenza held at the Royal Society of Medicine two days after the Armistice Newsholme had gone so far as to acknowledge that while the existence of a virus was still a 'moot point' there was no doubt that *H. influenzae* acting in 'conspiracy' with secondary bacteria such as streptococcus had greatly increased mortality from influenza. Nevertheless, while acknowledging that bacterial vaccines might eventually prove useful in preventing pneumonia, he thought it unlikely they would be much use against influenza so long as its aetiology remained a mystery. Warming to his theme, Newsholme had gone on to argue that while it had not been practical while war was raging to enforce measures designed to limit the spread of influenza it would be naive to think that such measures would prove any more effective now Britain was at peace. The only thing that could be said with certainty was that, as with the common cold, avoiding confined spaces and spending as much time as possible in the open air seemed to afford a degree of protection against flu. Douching of the nose and throat with an antiseptic could also be beneficial, he admitted, as could the use of 'veils' and respirators – though he doubted whether many people with catarrh would agree to submit to such regimens. However, he clearly had no intention of distributing face masks, though several speakers raised the issue. Nor was Newsholme willing to revive his plans, shelved the previous summer, for restrictions on public gatherings and travel. Yes the war was over, but were such measures really worth the 'heavy price,' he asked? It was a fair question. Tens of thousands of soldiers would soon be returning from the front, presenting a demanding logistical challenge, and wives desperate to be reunited with their husbands would be unlikely to tolerate delays. Having said that, if Newsholme cared to look abroad he would have found plenty of examples of alternative approaches.

In Australia, for instance, the authorities had imposed strict port quarantines, detaining more than 300 vessels and 80,000 passengers and crew between August and the end of November 1918. The measures had worked – up to a point. In Australia there had been no second wave that autumn, but in January, as Australian troops began returning from Europe, Melbourne suffered its first case of 'pneumonic influenza.' Other cases followed soon after in Sydney and by the end of the year some 12,000 Australians were dead.[149] Quarantines were rather

more successful in Samoa where the American governor mobi-
lized the indigenous population to keep foreign shipping out
with the result that not one person on the island contracted
the disease. By contrast, Western Samoa, where there were
no quarantines, was ravaged, with an estimated 22 per cent of
the population dying, one of the highest rates of mortality
from influenza anywhere in the world.[150]

However, it was in tough local measures adopted in the United
States that the contrast with British attitudes was most marked.
In San Francisco, for instance, schools, theatres and cinemas
were hurriedly closed and the municipal authorities issued an
ordinance mandating the wearing of gauze masks in public. As
public health notices on street corners reminded San Franciscans:
'Obey the laws/And wear the gauze/Protect yourself from septic
paws.'[151] Meanwhile, at the Naval Training Station on Yerba
Buena, home to some 4,000 cadets, the authorities imposed a
nine-week quarantine during which time drinking fountains were
sterilized hourly with blow torches and trainees required to keep
within 20 feet apart.[152]

Other US cities took similarly drastic actions. In Chicago and
New York the police were ordered to fine anyone caught
sneezing in public. Congress closed the public galleries in
the Senate and the House, while in the streets surrounding
the Capitol it became an offence for the sick to leave their
homes. Elsewhere citizens were urged to 'abandon the universal practice of shaking hands,' while Boston had churchless
Sundays.[153]

Only Philadelphia, it seems, failed to take adequate precautions. Perhaps it was because Philadelphia was a military
centre and the city was in the grip of war fever. Whatever the
case, on 28 September the Philadelphian authorities decided
to ignore evidence that influenza had returned to the eastern
seaboard and go ahead with a scheduled Liberty Loan Drive.

It was a decision they would regret. As in Manchester on Armistice Day tens of thousands of people flocked to the city centre. Within two weeks Philadelphia had recorded more than 2,600 deaths and by the third week of October deaths had soared to over 4,500.[154]

Though the end of war and the waning of the second wave presented Newsholme with an opportunity to mimic the strict hygiene and social distancing measures adopted in San Francisco, Chicago and New York, one could argue that he also flunked the test. Another half-hearted memo was issued, this time to the Royal College of Surgeons reminding their members of the LGB's desire for more data on the outbreaks. But, aside from insisting on stricter ventilation of cinemas, the LGB's only other initiative was to distribute a 15-minute film entitled, *Dr Wise on Influenza*. Prefaced by an appeal from the president of the LGB Sir Auckland Geddes this consisted of a lecture by a doctor explaining the precautions for avoiding influenza accompanied by 'cinematographic photographs in a popular vein.' Unfortunately, the LGB only had limited copies and although Manchester succeeded in getting hold of the film many authorities did not. As had been the case at the height of the autumn wave, Newsholme's critics in the popular press were unimpressed. *Punch* magazine was particularly scathing, satirizing the LGB's complacency thus: '"Our chief hope of control of influenza," writes Sir Arthur Newsholme of the Local Government, "lies in further investigation." Persons who insist on having influenza between now and Easter will do so at their own risk.'[155]

***

In mid November, for all that the second wave had yet to peak in Manchester, in London and other British cities the worst appeared to be over and with the news that Germany had signed the Armistice the nation erupted in a riot of celebration accompanied by patriotic flag-waving. But though the Armistice had removed the most immediate threat to life and limb, the influenza had not gone away, and as Lloyd George's new government began the slow and laborious process of demobilization Britons can hardly have suspected that the returning troops were now the carriers of a deadly new threat.

On the day after the Armistice soldiers had poured into the capital to join the munitions girls and office workers celebrating in the streets, swelling the crowds to such an extent that journalists observing the scenes reported that the predominant colour was khaki. There was then a delay as the government put in place procedures for demobilization in an 'orderly fashion,' with priority being given to men employed in pivotal industries such as coal-mining, many of whom had only been called up in the latter stages of the war. However, by December many soldiers were understandably becoming impatient to be reunited with their families and there was a growing sense of unfairness that those with long-service, who had endured and sacrificed the most, should be at the back of the queue. Following the eruption of small-scale mutinies at camps in Calais and Folkestone and a demonstration by 3,000 soldiers in London, in January the new War Minister, Winston Churchill, announced a new and more equitable scheme based on length of service and the number of times a man had been wounded in battle. The result was that by late January upwards of 600,000 soldiers had been demobilized and by early February men were returning to Britain at the rate of about 100,000 a week.

The first intimation that the flu still had the power to claim the lives of men and women in the prime of life came on new year's day when several newspapers splashed with the news that Captain Leefe Robinson, the first airman to bring down a Zeppelin on English soil, had succumbed to the virus. Robinson, who was 23, had only just returned to England, having been taken prisoner by the Germans in April 1917. During his incarceration he had made several attempts to escape from captivity, prompting the Commandant of the POW Camp at Holzminden where Robinson was being held to place him in solitary confinement. Robinson's health had suffered accordingly and on landing at Leith on 14 December many people were shocked to see this gallant airman and recipient of the Victoria Cross walking with the aid of a crutch.

On 23 December Robinson travelled to Harrow Weald in north London, not far from the site where in September 1916 he had flown alongside a massive SL11 German airship and riddled its hull with bullets, to spend Christmas with his family. But shortly after arriving he contracted influenza and on New Year's Eve he slipped into a delirium. According to his fiancé who nursed him, on several occasions he called out, imagining that the German sentries were standing over his bed with fixed bayonets, before eventually expiring of heart failure later that night.

Robinson's funeral three days later was front-page news. Hundreds of well-wishers gathered to watch the procession and as a large wreath was placed on the coffin a squadron of aircraft overflew his house. Robinson's coffin was then borne on an Air Force trailer to the tiny Harrow Weald cemetery where he was buried in an intimate private ceremony attended only by family and close friends. It was, reported the *Daily Sketch*, a fitting send off for a man who always bore his honor 'unassumingly.'[156]

A week later *The Times's* Rome correspondent reported that Italy was in the grip of a 'third wave of the epidemic' and on 16 January the paper reported that a 'giant' in Nottingham had died of influenza – the giant in question being Albert Brough, the seven-feet, seven-and-half-inch high proprietor of the Cremorne Hotel. Apparently, Brough's coffin was so long that there was no hearse in Nottingham large enough to transport it and a vehicle had to be altered especially.[157]

By the end of the month, with temperatures plunging and snow turning pavements in the capital slushy, *The Times* reported the country was in the grip of a new wave of 'considerable virulence' and advised people to go out of their way to avoid contact with the sick and, if impossible, to wear gauze masks or handkerchiefs over their faces. However, it was not until the week ending 8 February that deaths from influenza and pneumonia began to creep ominously upwards in London with, once again, the boroughs with major train terminuses showing the first distinct rises. A week later deaths from flu in the capital were running at 653 a week, the highest level since 7 December. By now flu had also returned to Liverpool, Newcastle, Bradford and Manchester, and on 18 February, beneath a headline declaring that the Germans had acceded to the Allies terms for a continuance of the armistice, *The Times* reported that Glasgow was once again badly affected with 141 deaths.[158]

The role played by the return of men from the Front in the recrudescence of influenza in the winter of 1919 is nowhere better illustrated than by the story of Robert Graves. Like many men awaiting demobilization in February 1919 Graves was anxious to be reunited with his family and growing increasingly impatient at the time it was taking. No doubt his annoyance was exacerbated by his sense of destiny. A close friend of Siegfried Sassoon, Graves had used the war to establish a reputation as a poet. Like Sassoon he'd also been severely wounded in 1917 at the Somme and suffered badly from shell shock, though unlike Sassoon Graves had never been hospitalized for the condition.

While on leave in Ireland, Graves had received a telegram from the War Office confirming his papers had come through. But in order to get the required stamp and be passed out he first had to return to the demobilization centre near his parent's home in Wimbledon, south London. Unfortunately for Graves, the demobilization of troops in Ireland had just been suspended due to the 'Troubles.' To complicate matters, Graves also had 'the beginning of influenza.' Fearing for his lungs if he was sent to an Irish hospital Graves decided to make a run for it. Gathering up his kit bag, he got an orderly room sergeant to make out his travel papers and hopped on the 6.15pm train from Limerick. Although Graves left without the proper demobilization 'code-marks' he didn't care as long as reached London. As he put it in his subsequent memoir *Goodbye to All That*: 'Wimbledon was in England and I would at least have my influenza out in an English and not an Irish hospital.'[159]

On the ferry to Fishguard on the night of 13 February his temperature ran high. Nevertheless, his mind was working clearly 'as it always does in a fever,' and on discovering the next morning that a strike was threatened on the London Electric Railways Graves decided to hop straight on a train from Wales to Paddington in the hope of being able to beat the disruption and catch a connecting train from Waterloo to his home in Hove. Graves was in luck. Not only were the steam trains from Waterloo still running, but on arriving at Paddington he found himself sharing a taxi with a stranded soldier who just happened to be the Cork District Demobilization Officer and who agreed to provide him with the missing code-marks. On reaching the Wimbledon demobilization centre, Graves had another lucky break: as he had just come from Ireland and Ireland was still officially a 'theatre of war' he was given priority over the other men waiting in line and was passed out on the spot. He just had time to pay a visit to his parents nearby before catching a train from Waterloo, arriving at Hove just as his wife and mother-in-law were sitting down to supper. By now, however, Graves was severely ill. It was like 'a bad dream,' he writes. 'I did not know what to do. I knew I was ill, but this was worse than illness.'

Warning them he had influenza, Graves put himself to bed. Within two days nearly everyone in the house, with the exception of Grave's father-in-law and a servant who kept a gypsy's charm around her neck, were also sick. Brighton, like other towns in Britain, was still suffering from an acute shortage of nurses and doctors but eventually a docor was found. On examining Graves he had bad news: he had septic pneumonia in both lungs and had 'no chance of recovering.'

Graves, who is not averse to playing fast and loose with the facts, may have been exaggerating the gravity of his disease for narrative effect – or he may really have been as ill as he makes out. Whatever, he made a miraculous recovery, writing that 'having come through the war, I would not allow myself to succumb to Spanish influenza.' Within a few weeks, he even claims he was sufficiently recovered to witness soldiers from a local Army camp parading through the streets of Brighton in protest at the glacial-like speed of demobilization.

Reading Graves' account one could be forgiven for thinking that the third wave of influenza was an amusing diversion, and certainly nothing like as fearful as the autumn wave or anything that Graves, or men like him, had witnessed at the Front. And certainly to judge by the mortality figures, the third wave *was* far less fatal. In London some 6,000 people were to die from flu and related pneumonias between February and May 1919, compared to nearly 16,000 in the autumn. And for the country as a whole the third wave accounted for just a quarter of the overall mortality from the pandemic – the vast majority of the deaths (64 per cent) being concentrated in the period September–December 1918.[160]

But for all that the obituary columns in *The Times* may have been less extensive, for those in the front line of the successive waves of influenza the new onslaught was just as grim. In a series of notebooks compiled in the 1950's, Dr Basil Hood, the medical superintendent in charge of St Marylebone Infirmary in north Kensington, recorded how the hospital had 'literally reeled' under the repeated outbreaks. 'The epidemic was certainly the worst and most distressing occurrence of my professional life,' he writes. 'Our helplessness now 30 years later would be nearly as great...'[161]

According to Hood, his ordeal had begun in the spring of 1917 when the War Office had issued an appeal for doctors and nurses to serve at the front, leaving him with a small skeleton staff. The result was that when in October 1918 200 sick soldiers were suddenly transferred to St Marylebone from a military infirmary in Paddington, Hood was ill-equipped to deal with the epidemic that broke upon his wards.

All training and indeed every sort of trimming went by the board whilst the staff fought like Trojans to feed the patients, scramble as best they could through the most elementary nursing and keep the delirious in bed. Each day the difficulties became more pronounced as the patients increased and the nurses decreased, going down like ninepins themselves. Sad to relate some of these gallant girls lost their lives in this scene of never-to-be-forgotten scourge...as I write I can see them now literally fighting to save their friends, then... dying themselves.

Hood recalled the fate of one nurse in particular who contracted 'a terribly acute influenzal pneumonia' and insisted on being propped up against the wall by her bed until she drowned in 'her profuse, thin bloodstained sputum.' In the case of another nurse, Sister Walker, there was nothing Hood could say or do to persuade her to take fewer risks, particularly when a colleague fell ill and she insisted on nursing her.

'I can see her now buttonholing me in the passage, "Did I think sister was a little better?" "Was she doing well?" "Was there anything more she could do?" In the end both the woman she was nursing and Sister Walker died. 'It was a real blow when she was laid to rest,' writes Hood. 'I have never really got[ten] over that time and no wonder.'

Amid the deaths, however, there had also been moments of levity. At one point there was such pressure on beds that Hood decided to charter a double-decker bus to return recovered patients to their homes. Hood drew up a list of the patients, together with their addresses and a plan for dropping them off in 'a suitable order geographically.' Unfortunately, the porter who Hood selected for the task became confused and delivered 'the wrong wives to the wrong husbands' only to realize his mistake and have to 'sheepishly' retrace his steps.

At the end of November the stress and strain on Hood proved too much – he had been working 15 hour days as well as nights – and he collapsed with fatigue ('I could barely stand and always when possible against a wall,' he writes). Hood spent the next three months convalescing at home, returning in mid-February just in time to see the third wave break upon the wards. At that time North Kensington was an area of slums and the hospital was soon filled with the local urban poor. Although the war had been over for nearly three months, Hood writes that these people were still badly 'debilitated and under-nourished' and did not enjoy the same resistance that 'more fortunate people had.' As before, the most distressing cases were those with influenzal pneumonia and women in the later stages of pregnancy. There were also further deaths of nurses who selflessly put themselves in harm's way.

From Hood's notes, it is difficult to work out just how many patients he treated during this final wave. At the height of the autumn outbreak, there had been some 800 patients on the wards. By the end of the third wave, Hood had admitted perhaps a hundred more, although by now he was also able to call on Chelsea infirmary, which agreed to take 40 cases that would otherwise have come to him. The final tally in his notebook reads '850 influenzal cases' of which 364 were suffering with 'influenzal pneumonia' and '197 died (8 being pregnant).' However, in a further note dated 2 April 1919, Hood has corrected the tally to read 407 cases of acute pneumonia 'of which 208 died and 199 recovered.'

The grim reality behind these figures is nowhere better illustrated than by the experience of Bertram Copping, the newsagent's son from Islington. After the doctor's initial visit to his father's bedside, Copping had made repeated efforts to get him to return, but by the time he did so it was too late and his father was dead. By now Copping's mother was also ill so he was sent to make the funeral arrangements and select a coffin. 'That a small boy should come on such an errand caused no surprise to the undertaker – by that time he was used to it, and he patiently took down all the details of time and place,' reports Copping.

Next, it fell to Copping to choose the coffin. No one had told him how much he could spend and at first he was baffled by the choice of models. Eventually, fearful of what his mother would say if he selected the cheapest, he opted for the next one up in price – '"the cheapest but one" being a guiding principle in my early years as a young man.' When the day of the funeral arrived there were so many floral tributes from friends and customers that they spilled out of the house onto the pavement. But the image that haunted Copping for the rest of his life was the scene at the church. When he buried his mother in 1954, Copping writes there were just three coffins at the service. But when he buried his father in February 1919 the coffins were stacked 'one on top of the other' and there were so many mourners that the church was full to over-flowing. 'It was at that moment at the age of ten that my boyhood ended.'[162]

Copping was not the only person for whom the epidemic marked a loss of innocence. Anthony Burgess, the future author of such diverse books as *A Clockwork Orange* and *A Mouthful of Air*, was aged just one when in February 1919 influenza visited his home in Harpburhey in northeast Manchester. In his memoir, *Little Wilson and Big God*, Burgess recalls how the virus claimed the lives of both his mother and sister while he lay 'apparently chuckling' in his cot. According to Burgess his father had yet to be demobilized and had discovered their bodies by chance on one of his irregular visits to the family home in Carisbrook Street.

By now, Niven was only too aware of how easily the virus was passed between people. In the third week of February, as deaths from influenza and pneumonia in Manchester approached the 200 mark, he had 50,000 more handbills printed reiterating his warnings about the disease being communicated from the sick to the healthy via 'coughing, sneezing and speaking.' Niven also took up the growing calls in the press for the adop-tion of muslin face masks and for people to use handkerchiefs to cover their mouths. Next, having alerted the police of the need once more to enforce the LGB's edict restricting the opening times of cinemas, Niven turned his attention to other places where people liked to gather in large numbers, such as pubs. During an inspection of ale houses in December, Niven had noticed how publicans frequently re-used beer glasses without bothering to wash them. Coupled with the numbers of people crowded together in such places, Niven argued that this created the risk of the disease being passed between customers. Meanwhile, in the home he thought he'd spotted another common source of infection: the family loaf. He lectured that it was no use family members keeping their distance if those infected, as well as those waiting on them, placed their hands on bread without cleaning them. 'Any one handling food such as bread, &c., should wash their hands carefully before doing so.'

Finally, as the cold weather continued to bite, Niven urged Mancunians to keep themselves well fed and warm, if necessary by wearing extra underclothing and good boots. 'The disease not only infects more easily, but is more severe, where deficiency of food, clothing and fuel is present,' he warned.

***

Following his recovery from influenza in October, Lloyd George had been too busy shuttling back and forth between London and Versailles to dwell on his illness let alone the possibility of a general recrudescence of the disease. Then in December had come the general election followed by a series of meetings with Clemenceau and Wilson over the forthcoming peace talks in Paris. The US president had staked his political reputation on being able to deliver a just and equitable settlement that wouldn't overly punish the German people and would ensure the impartial settlement of colonial claims while guaranteeing freedom of the seas. However, the British and French, while paying lip service to Wilson's 'Fourteen Points,' had different ideas. As Sir Eric Campbell-Geddes, Lloyd George's Minister without Portfolio, had put it in a speech in Cambridge shortly before the general election, 'The Germans... are going to be squeezed, as a lemon is squeezed – until the pip's squeak.'

The stage was thus set for a confrontation between the American desire for a less rancorous and freer new world and the desire of Britain, France and Italy for revenge and compensation for the war. Negotiating a settlement acceptable to each member of the Big Four would take all of Lloyd George's political skills. What he couldn't know is that the Paris peace talks would also take place in what one member of the American delegation called a 'fug of flu.'

In the early winter of 1919 Paris, like London, was a city still reeling from the after-effects of war and influenza. The autumn wave had exacted a heavy toll on both soldiers and civilians. In September some 25,000 men in the French armed forces were laid low by flu and by the last week of the month the disease was killing some 30 soldiers a day. During ten days in October, a further 36,000 French troops succumbed and 2,400 died, with those who had previously been gassed appearing particularly vulnerable.[163] In the French capital it is estimated that between early September and the middle of December, 19,000 Parisians died from all causes. What proportion of these deaths was due to the epidemic alone is hard to compute as only flu deaths that came to the attention of the state's *médecins-inspecteurs* were recorded as such. Nonetheless, in this period there were some 7,000 official influenza deaths and nearly 3,000 more from pneumonia.[164]

One of the most prominent victims was the avant-garde poet Guillaume Apollinaire. A naturalized Frenchman (he was born Wilhelm de Kostrowitsky in Rome of mixed Italian and Polish ancestry) Apollinaire had volunteered for the French Army in 1914 and in 1916 had received a serious shrapnel wound to the head. While being trepanned at the Italian hospital in Paris he published his first play and coined the term surrealism, but on 3 November 1918, while still recovering from his head injuries, he contracted a particularly virulent strain of flu. His friend Blaise Cendrars, who had recently lunched with him in Montparnasse, arrived at his flat to find the 38-year-old poet 'all black' from cyanosis. Cendrars went to get a doctor but he was too late and on 9 November Apollinaire died.

The demise of such a vigorous and romantic figure seemed to symbolize the disproportionate toll that the flu exacted on men and women in the prime of life and soon those of a similar age attacked by the disease were said to be suffering 'Apollinaire's syndrome.' To Cendrars, the fact that Apollinaire's death had come just two days before the Armistice struck him as absurd and leaving Père Lachaise cemetery, where he had gone to pay his last respects, he convinced himself that his friend's death was a joke and that Apollinaire would soon be resurrected. 'Under what mask will Guillaume return to the great celebration in Paris?' Cendrars asked rhetorically. But like tens of thousands of his adopted countrymen Apollinaire didn't come back. The father of surrealism was gone for ever.[165]

After Apollinaire's death, the autumn wave abated as it did in other European capitals. But though the death rate in Paris would never be as high again, that winter flu – and, just as important, respiratory diseases – never really went away. Some 1,400 Parisians died of flu and pneumonia in December and a further 1,500 in January. The last ten days of January were particularly bad, provoking one of President Wilson's aides to remark that there seemed to be 'millions of throat germs going round.' The casualties included several members of the American peace delegation, including Colonel Edward House, Wilson's special representative in Europe and a key member of his negotiating team. House had first contracted influenza in late November but towards the end of December he suffered a relapse and in January he fell sick again, this time with a kidney stone. House's protégé, Charles Seymour, was also out of sorts, complaining 'everyone has a cold here, and the Paris cold seems to take it out of one more than any I have known.' Flu also struck down another American aide, the Colombia University historian James T. Shotwell, while Wilson's principal financial adviser, Norman Davis, contracted pneumonia. The American historian Alfred Crosby argues that the illness of House and other key members of the American delegation was a severe blow to Wilson coming at a critical time in his pre-conference planning when he needed everyone to be at the top of their game in order to counteract British and French manoeuvring.[166] However, the Americans were not the only ones inconvenienced by the flu. By now several other diplomats at the talks were suffering 'colds' or had lost their voices. Then, 30 days after the opening of the conference, came the news of the death of a key member of Lloyd George's team.

A baronet and MP, Sir Mark Sykes had arrived in Paris on 4 February 1919 as a member of the British delegation that was negotiating the post-war partition of the Ottoman Empire. A close friend of T.E. Lawrence, Sykes was a leading Arab expert who in 1916 had helped draw up a secret Anglo-French agreement defining the respective spheres of influence of the Allied powers in the Middle East. On 30 October, just as the mortality from the autumn wave was mounting, Sykes was sent to Jerusalem to smooth the way for the planned post-war partition. The result was that he missed the Armistice and the worst weeks of the influenza. Nevertheless, when Sykes returned to Britain on 30 January, 1919, his wife, Edith, was shocked to find him thin and emaciated. Sykes informed her he had contracted a virus in Syria which prevented him keeping down solid food and for three weeks had been living on nothing but condensed milk. Edith pleaded with him to rest but, convinced his expertise was needed in Paris, Sykes insisted on travelling to the French capital. Edith decided to accompany him in the hope of keeping an eye on him, but no sooner had they taken a room at the Hotel Lotti near the Rue de Rivoli then she contracted flu and was forced to retire to bed. Sykes, however, rushed off to do the diplomatic rounds. This final burst of frenetic activity used up his reserves of energy and on 10 February, after a night at the opera, he collapsed, telling his secretary 'I've got it.'[167]

Edith immediately rose from her own sickbed to nurse him, ordering up supplies of Bovril, Oxo and various patent medicines. But it was no use and on 16 February Sykes died of pneumonia accompanied by an agonizing infection of the ear. He was buried two weeks later with full military honours in a lead lined coffin near his family seat at Sledmere, Yorkshire.[168]

Interestingly, when Lloyd George had gone to Manchester the previous September to deliver his speech at the Hippodrome Sykes had been in the audience, having travelled up from London with the Prime Minister on the same train. However, Sykes did not contract influenza that autumn and by November he was in the Middle East so it is possible that, unlike other members of the British delegation in Paris in February, he did not have any antibodies against the strain of influenza circulating in Europe. On the other hand, studies conducted after the pandemic found little evidence that attacks in either the summer or autumn conferred protection against the subsequent winter wave, so it is possible that the third wave of flu was sufficiently different antigenically from earlier versions of the virus that many people were unprotected. In other words, Sykes's death may simply have been bad luck.[169] Certainly, his illness coincided with the onset of the winter wave in Paris. That week 426 Parisians died of influenza and 513 died of pneumonia. As in London, the peak was reached in Paris in the week ending 1 March with deaths falling steadily thereafter.

There is one final coda to the story of the Parisian wave: the illness of Woodrow Wilson himself. On 3 April, with Wilson locked in tense negotiations with Clemenceau over the issue of reparations and Germany's new borders, the US president suddenly began coughing convulsively. By the evening he was running a temperature of 39° C and was seized with severe stomach cramps. That night the president's life hung in the balance. The diagnosis was influenza. Wilson's doctor blamed the infection on Clemenceau, who had been suffering with a cough for several weeks, but there is no evidence the French leader had flu and Wilson could just as easily have contracted the virus from a member of the American delegation.

Too ill to continue conducting the negotiations in person, Wilson appointed House as his go-between. It was a critical moment. By now, Wilson was so tired and disheartened by the lack of progress in the talks that he was seriously considering withdrawing from the conference. But that is not what happened. Instead, he spent the next five days in bed, rejoining the negotiating table on 8 April. But Wilson was not the same man he had been before his illness. Edmund Starling, his secret service guard, commented that the flu had left the President very weak so that he 'lacked his old quickness of grasp.' Others close to Wilson noted how his left eye drooped and the left side of his face twitched and Gilbert Close, his personal secretary, felt that while laying in bed the president had manifested mental 'peculiarities.' Writing some time after the conference Herbert Hoover also felt that Wilson's illness had robbed him of his former incisiveness and mental acuity, recalling that 'others as well as I found we had to push against an unwilling mind.'[170]

Once again, it is fascinating to speculate on what part influenza played in the final Versailles agreement. Did Wilson's illness contribute to the breaking of the Franco-American stalemate on reparations and the occupation of the Rhineland and the Saar? Or would Wilson, encumbered by an isolationist Senate and an American domestic public hostile to any breaches with the Allies, have been forced to compromise his principles sooner or later anyway? Crosby makes a strong case for the former and certainly by the time Wilson rejoined the nego-tiating table on 8 April the talks were moving decidedly in favour of the French and British position. However that may have had as much to do with Clemenceau's intransigence and Lloyd George's playing off of one personality against another. Nor can Wilson's illness be blamed for the insertion of the war guilt clause and France's insistence on an open-ended commit-ment for German reparations. A better case can be made for the connection between Wilson's illness and his failure the following September to persuade the Senate to ratify the Versailles Treaty and commit the US to the League of Nations. But by then the president had suffered a major stroke and had virtually withdrawn from public life.[171]

***

Just as it is supposed that diseases can influence historical events so it is sometimes argued that the responses to plagues can be mitigated by political and social conditions and the influence, for want of a better phrase, of 'national character-istics.' In 1918 at the height of the winter wave *The Times* made precisely this argument asserting that the German air raids had been a 'summer shower' compared to the deluge of influenza germs Londoners had just received and that 'never, perhaps, has a plague been more stoically accepted.'[172]

But like the argument that the flu influenced the outcome of the peace the notion of British *sang froid* in face of influenza's depredations may also be a myth. True, Britons did not flee to the countryside as they had during earlier 19th century cholera outbreaks, but this was arguably because influenza, though widespread, was far less likely to kill (and when it did, the corpse was less likely to be viewed as an object of revulsion). Nor did Britons panic at the sight of strangers suddenly collapsing in the streets or of loved ones turning blue as the virus made its way to their lungs. Perhaps they would have done if, as today, scientists had understood the pathogenesis of influenza and the reporting had been more sensational. But Britain was at war and the entire political machinery of government and civil society was geared to reinforcing the message that Britons had what it took – 'that subtle strength which we call staying power' – to defeat the Germans and other external threats to the nation. That was a message that neither the Zeppelin raids, nor the mounting food and fuel shortages, nor the deaths of tens of thousands of civilians from a mysterious foreign plague could be allowed to interfere with. Nevertheless, even before the Armistice, there were signs that the 'stiff upper lip' was more fragile than the official narrative allowed. In Catholic west Belfast where the writ of the British state was challenged by Irish nationalist sympathies, the autumn wave had sparked mini riots, and even in north London there had been runs on chemists and panicked queues outside doctor's surgeries. Then there were the far from stoical criticisms of the LGB's mishandling of the crisis and the National Service Board's delay in releasing doctors from the military. But perhaps the most revealing phenomenon of all was the spate of influenza-related suicides. These had come to public attention at the height of the winter wave, though with Britain still at war they had commanded little column space. Following the Armistice, however, such reports became more frequent and were given greater prominence.

In December, for instance, the *Manchester Evening News* reported that a gardener named Williams from Portmadoc, in

North Wales, had attacked his wife and six children with a razor while suffering the after effects of influenza. His family survived but he killed himself. In the next paragraph the paper reported that a woman named Ellen Booth had also committed suicide by slitting her throat with a butcher's knife. She too had been suffering influenza and had recently lost her husband to the disease.[173]

Then in January the *Hackney Gazette* carried a series of reports about the local suicides of young women. The victims included a 23-year-old who was due to be married but felt she'd 'never be well again' and didn't want to be a 'burden' on her family and fiancé. By April there were also reports of doctors committing suicide following episodes of flu.[174] But perhaps the best example was the suicide of Lance Corporal James Ernest Jones of the Royal Engineers – what the *Observer* billed as a 'remarkable story of delusions and despair.'[175] At the inquest, the coroner heard that Jones had first tried to kill himself in February while awaiting demobilization (he'd apparently been under the delusion that he had a degenerative disease and was about to be placed in a lethal gas chamber). After being admitted to Stoke Hospital, doctors assigned him light duties and arranged for a non-commissioned officer to watch over him. But one night Jones escaped. Nothing further was then heard of him until he turned up at his mother's house in March saying he was about to be demobilized. He spent two weeks at his mother's but when two Grenadier Guardsmen arrived to arrest him he disappeared. By the time Jones was discovered again it was May. He was immediately taken into custody and stripped of his uniform. However, on 27 May he escaped for a third time, returning to his mother's house where he pleaded for her to take him in. That night she gave him some food and put him to bed. The next morning she found him lying dead on the floor, his face pressed hard up against the gas nipple of the fire.

Ever since the 1890s it had been recognized that influenza could result in nervous complications and various nonspecific mental disturbances, including 'melancholy,' 'lassitude,' 'lethargy,' 'neurasthenia,' and 'neuritis.' In 1919 Sir George Savage, a consultant physician at Earlswood Asylum and an expert on mental disease, lumped these disturbances together under the heading 'psychoses of influenza.' Pointing out that in both the Russian flu pandemic and the 1918–19 outbreak there had been a considerable number of cases of 'mania' and 'insanity,' Savage argued that influenza epidemics fell with 'greatest potency upon the nervous system' and that patients admitted to asylums during such epidemics 'have always to be considered as potentially suicidal.' Other experts such as Wilfred Harris, physician to the Hospital for Epilepsy and Paralysis, in Maida Vale, went further arguing that no other disease was associated with such frequent and varied nervous sequelae and that 'the prime sufferer in influenza is the nervous system,' while Claye Shaw, a lecturer on psychological medicine at St Bartholomew's Hospital, argued that influenza 'even in the most favourable cases' is usually accompanied by depression.'[176]

Today, the association between influenza and complications of the central nervous system (CNS) is well recognized even if the pathogenic mechanisms are still not fully understood. CNS complications of influenza include Reye's syndrome, acute necrotizing encephalopathy, myelitis and auto-immune conditions such as Guillaume-Barre's syndrome. In addition, influenza has been associated with a range of neurological conditions, including schizophrenia, epidemic encephalitis, affective psychosis and Parkinson's disease. But perhaps the most controversial sequelae of influenza is Economo's disease, the 'sleeping sickness' which first appeared in France and Britain around 1917 and was endemic to Europe and America from 1919 to 1929. Economo's disease, also known as enceph-

THE PREVAILING EPIDEMIC.

"Ah! You may laugh, my Boy; but it's no joke being Funny with the Influenza!"

Figure 5.3 Mr Punch wrapped up in blankets in front of the fire, eating gruel and suffering from influenza. The caption reads: 'Ah! You may laugh, my boy; but it's no joke being funny with the influenza!' Credit: Wellcome Library, London

alitis lethargica, was unusually devastating because it acted in two phases, first causing a brain damage that left its victims in a statue-like condition in which they could neither move nor speak, then, after a symptom-free interval of many years, producing a form of Parkinson's disease in about 80 per cent of survivors. Because its appearance more or less coincided with the influenza pandemic, many contemporary epidemiologists were convinced there was an association. However, both Baron Constantin von Economo, the neurologist who lent his name to the disease, and Dr Oliver Sacks, whose 1973 book *Awakenings* about reviving encephalitis patients with the miracle drug L-DOPA, had their doubts and it is only recently that evidence has been adduced showing that the supposition of a link was probably correct.[177] The debate is of more than theoretical interest for if it could be shown that influenza and Economo's disease were causally related then that would add a further five million deaths to the pandemic's toll. Although the suicides that followed the influenza were probably too few in number to make much difference to the overall tally, they are significant for another reason. For while, as with Economo's disease, it is impossible to demonstrate a causal relationship between influenza and suicide, the timing of the cases is hard to ignore and suggests that for all the stoicism evinced by *The Times* the pandemic provoked a range of responses, including depression and a despair so absolute that many survivors were driven to take their own lives.

***

The third wave of the Spanish influenza was to be the last. By the time infections abated in May the virus had killed a further 4,000 Parisians and an additional 5,000 Londoners. In Manchester, fully a third of the 3,000 victims of the pandemic

– or some 1,000 people – died in a 12-week period between 8 February and 26 April 1919. In all, the three British waves of the disease lasted just 46 weeks and killed an estimated 228,000 people. That was an extraordinary number of victims to be concentrated in such a short period of time. During the Russian flu the second wave had peaked in spring 1891, more than a year after the first wave. The compression of three waves into an 11-month period suggested that either the surface proteins of the 1918–19 virus drifted more rapidly than previous pandemic strains or that it had an unusually effective mechanism for evading the human immune system. The other unusual aspect was the high case mortality rate – 2.5 per cent compared to 0.1 per cent in a normal influenza epidemic – and the fact that fully half the victims were young adults aged 20–40. This was the complete reverse of the Russian flu in which mortality was highest in the over 65s and children under five and which produced the usual U-shaped mortality curve. By contrast the graphs of mortality in the 1918–19 pandemic were W-shaped, reflecting the unusual peak in deaths in middle life.

In the US, where the pandemic killed an estimated 675,000 Americans and reduced life expectancy by ten years, the influenza showed a similar proclivity for the fittest and healthiest members of the population, scotching theories that the W-shaped curve observed in Britain and France was related to the deprivations of a wartime economy. No doubt the movement of troops and the crowding of civilians had played a part, argued Britain's Chief Medical Officer George Newman, but 'we cannot escape the knowledge that the epidemic ravaged populations of this age period who were not subjected to these unfavourable circumstances.'[178]

Moreover, if the mortality had been bad in Britain it was even worse in distant outposts of the British empire. In South

Africa, for instance, some 140,000 people perished, the vast majority non-Europeans. Influenza had arrived in Durban as early as 14 September. From there it spread to the central Rand decimating black labourers who worked in the gold mines and by the 23rd of the month it had reached Kimberley. However, it was in Cape Town, where the flu claimed the lives of some 6,000 citizens in the autumn, that its depreda- tions were most keenly felt. The first outbreaks had occurred among Nigerian troops and the South African Native Labour Contingent just returned from France. Next, servants at the city's hospitals fell ill, followed by porters and ward maids. Soon deaths were running at well over 500 a day and within two weeks Cape Town 'was like a city of the dead.'[179] Covered wagons were dispatched to collect the bodies of people who had collapsed in the streets while health visitors were sent door-to-door to check on families. The head sister of a leading Cape Town hospital vividly recalled the scenes.

> …whole families stricken, the dead and living in the same beds, no food in the house, no one able to crawl about to get it; hundreds of people starving because they could not go out to get food; all delivery carts stopped, no one to drive them, shops shut, the people being ill; business houses shut up; trains and trams stopped running; theatres, bioscopes, and churches all empty and closed…[180]

In the cemetery six miles outside of Cape Town people had had to bury their relatives with their bare hands. 'Often they were so weak that they could only dig two or three feet deep, and as they turned to get the body they had brought other people came and threw the bodies of their friends into the grave.' The subsequent fights, she added, were 'terrible.'

Another man, writing to his mother in England, was similarly appalled. 'I have seen sights I have never seen before & never want to see again, of motor lorries loaded up with dead, carts of all sorts taking bodies to the Cemetery & from people carrying their dead or wheeling them on wheelbarrows with friends and relatives or helpers following with spades and pick-axes to dig the graves.' In Cape Town alone, he said, the influenza had made orphans of between two to three thousand children.[181]

However, it was in Asia and in India especially, where famine and the absence of developed medical care systems magnified the impact of the virus, that influenza was to claim the most lives. In Bombay, where the disease arrived courtesy of a container ship in May, it was reported that one million people had perished. In the Indian Army mortality rates were as high as 22 per cent while in the Punjab, one of the worst affected regions, streets and lanes were littered with the bodies of the dead and dying. In 1920 the Ministry of Health estimated the death toll in British India at five million, with perhaps a further one million deaths in the native states, giving a total death toll for the Indian subcontinent of six million. In 1927, that estimate was revised upwards to 12.5 million. The latest estimate is 18.5 million.[182]

A death of a friend or a member of the family hits us in the guts, but deaths of this order of magnitude, of people we have never seen and will never meet, are impossible to imagine and evoke little emotion. The scale is too vast, too epic. 'When one has fought a war, one hardly knows what a dead person is,' remarks Albert Camus. 'And if a dead man has no significance unless one has seen him dead, a hundred million bodies spread through history are just a mist drifting through the imagination.'[183]

Camus's reflections were sparked by an imaginary plague in Oman, in Algeria, but they apply equally to the 1918–19 influenza – perhaps more so. One of the curiosities of the Spanish influenza is how few people bothered to record their experiences for posterity – indeed, were it not for the deaths of prominent figures like Sir Mark Sykes and Apollinaire, or the letters collected by Richard Collier, we would be hard pressed to produce evidence the pandemic had occurred at all. 'The Spanish Lady inspired no songs, no legends, no work of art,' writes Collier. 'Even fundamental facts were meagre.'[184]

Reflecting on his experiences as an epidemiologist in the 1930s, Major Greenwood wrote that historians would have to go back to the bubonic plague of 1348–49 to find a disease event on a similar scale and suggested that it was because influenza lacked the 'emotional colour' of the plague that it had failed to register a similar impression.[185] And certainly it is noticeable that those artists, such as Robert Graves, who did write about the influenza, appear to have gone out of their way not to paint it in cataclysmic tones. Partly, this is a reflection of Graves's aesthetic sensibility but it also reflects what the literary critic Paul Fussell, who made a special study of the role of memory in the memoirs and poetry of WWI, calls 'the paradigm of ironic action.'

The Great War was so horrific, he argues, the loss of life so senseless and so contrary to the myths of heroism that had gone before, that irony was the only response that made sense. 'By applying to the past a paradigm of ironic action,' Fussell writes, 'a rememberer is enabled to locate, draw forth, and finally shape into significance an event or a moment which otherwise would merge without meaning into the general undifferentiated stream.'[186]

But if the war was meaningless it could not be ignored. Influenza, on the other hand, could and for the most part was (it was only later that medical historians piqued by this lack of parity appended the adjective 'Great' to influenza). In fact, faced with such huge numbers, amnesia may have been the psychological strategy that made most sense. As the American critic H.L. Mencken put it in 1956, reflecting on why Americans of his generation so seldom spoke of the pandemic: 'The human mind always tries to expunge the intolerable from memory.'

It took not only a great artist but one with a peculiar preoccupation with disease to look the Spanish influenza in the face and capture it in the vivid tones it deserved. The artist was Edvard Munch and it is in his two self-portraits painted in 1919 depicting his own illness and its after effects that we come closest to understanding what it must have been like to have suffered and survived the Great Flu.

Influenza was a subject made for Munch. One of the Norwegian artist's earliest memories was of his mother, confined with tuberculosis, gazing wistfully at the fields that stretched outside the window of their house in Kristiania (now Oslo). Following her death in 1868, Munch was left in the care of his father, Christian, a deeply religious doctor who opted to bring his son on house calls with him, thus deepening Munch's early acquaintance with illness and suffering. But it was his beloved elder sister Sophie's death from tuberculosis nine years later at the age of 15 that left the deepest scars. Dying, she asked to be lifted out of bed and placed in a chair. Munch, who was just one year younger than his sister, painted her sickroom obsessively, his many compositions reflecting Sophie's terminal illness and his shock at witnessing her death at such a young age.

In his *Self-Portrait With Spanish Flu*, painted in 1919, Munch sits in a chair just as Sophie had. Facing the viewer, with peering eyes and his mouth ajar, he looks exhausted and pale. His hands rest weakly on a red blanket folded over his lap, the lively colours of the blanket and the yellow of the chair contrasting markedly with his own lifelessness. The sense of debilitation and decay is emphasized by the flatness of the composition and the thin application of paint in which some critics have discerned a sense of disintegration. Only the crumpled duvet lying nearby and the fact that Munch is sitting upright rather than lying prone on his sickbed suggests he may be over the worst.

In the second painting, *Self-Portrait After Spanish Flu*, however, one has the impression that death may have been more merciful. Haggard and unshaven, Munch looms from the canvas his shoulders sagging beneath a suit that appears two sizes too big for him. His sunken eyes stare unblinking from a face etched with garish reds and blacks, while his chin and cheeks are outlined in a cool green, suggesting the residua of 'grippe' that still clings to him. It is the artist's eyes that draw you back, however. There is something unsettling in his stare, as if Munch has been through an unforgettable experience.

Looking at Munch's tortured visage one cannot help but be reminded that the Great Flu was very nearly forgotten. It is only his genius, and the extraordinary acts of memorializing by hundreds like him, that has kept the experience alive for future generations.

PART III

'We can't make this pandemic go away, because it's a
natural phenomenon, it will come. But what we can do
is to limit its impact.'

– Sir Liam Donaldson, Chief Medical Officer,
UK Department of Health[187]

6

Vietnam, February 2005

In an isolation ward of Hanoi's Bach Mai hospital a brother and sister are fighting for survival. Nguyen Sy Tuan, who is 21 and pencil-thin, appears to be having the worst of it. Five days ago he was rushed to Bach Mai from a provincial hospital 100 miles southeast of Hanoi complaining of a fever and chest pains. The only thing now keeping him alive is the respirator beside his bed. In the next cot along, beneath a sign marked 'quarantine,' lies his 14-year-old sister. Her infection was diagnosed quicker and so far her X-rays do not show the tell-tale shadows that would indicate the virus has burrowed deep into her lungs. Doctors are crossing their fingers they won't need to intubate. Instead, they hope to clear the teenager's infection with the aid of antibiotics and anti-viral drugs.

'We'd heard about bird flu, but we never imagined it would infect our family,' their older sister, Nhung Ngoan, tells me in the corridor outside their room where we are both dressed in identical white bio-contamination suits and blue surgical masks. 'We are really scared because we know it is a very serious disease.' Then, grasping her mask closer to her mouth for protection, she adds: 'All we can do is pray they will recover.'

Since the current strain of H5N1 emerged in Vietnam in late 2003, Nhung Ngoan is not the only one who has been praying. Looking at Vietnam and other countries in Southeast

Asia where people have been infected with bird flu, the question on the world's lips is could 1918 reoccur, could it happen again? Since the 1918 pandemic there have been two major 'shifts:' the 1957 'Asia flu' and the 1968 'Hong Kong' flu, both of which killed about one million people worldwide. Although the gap between the 1918 flu and the Asian flu was 39 years, on average flu pandemics occur every 27 years. In other words, the 20th century should have seen a fourth pandemic in 1995.

Pandemics are the viral equivalent of perfect storms. In order to trigger a shift on the scale of 1918 three things need to happen. First, a new influenza virus – one against which people have no or few antibodies – has to emerge from a 'hidden' animal reservoir, and second, the virus has to make people sick. Both of these events occurred in 1997 when a strain of H5N1 simultaneously infected poultry and humans in Hong Kong (in all 18 people were hospitalized and six died, prompting the Hong Kong authorities to cull 1.5 million chickens). The third thing that needs to happen is that the virus must be able to spread efficiently between people – preferably via a cough, sneeze or handshake. The current H5N1 subtype, like the 1997 Hong Kong strain from which it is descended, is highly pathogenic in chickens and at the time of writing there have been 385 human cases and 243 deaths, so it is safe to say that for some people H5N1 is both infectious and deadly. Whether or not the third condition is also on the point of being fulfilled depends in part on how you interpret the infection of Sy Tuan and his sister.

Sy Tuan's ordeal began on 8 February 2005, the eve of Tet, the Vietnamese holiday which marks the transition from one lunar year to another, when he visited a neighbour and bought a duck and a few chickens for the family pot. With the help of his sister Sy Tuan slaughtered the duck and poured its blood into a bowl. After adding vinegar to stop the blood congealing, Sy Tuan then set about preparing duck's blood soup, the traditional eve-of-Tet meal.

That evening Sy Tuan, his sister and their mother and father sat down to consume several bowls. Within five days Sy Tuan was so ill he couldn't move. Nevertheless, it was only after he'd been in bed for ten days that his father thought to take him to the local hospital in Thai Binh. By the time Sy Tuan arrived in Hanoi a few days later his condition was critical. Tests confirmed he had H5N1. The virus had apparently prompted Sy Tuan's immune system to go into overdrive, triggering a exuberant defensive reaction known as a 'cytokine storm.' The cytokines had swamped his lungs causing them to fill with fluids and damaged cell tissue which in turn had encouraged opportunistic bacterial infections. As Dr Nguyen Tuong Van, the director of the Bach Mai hospital's intensive care unit, told me: 'When I examined his chest X-ray there were white shadows everywhere. It was like looking at a patient in the advanced stages of HIV.'[188]

A few days later Ngoan also fell ill and was rushed to Hanoi with a temperature of 40° C. Tests showed she also had H5N1. Then, on 27 February a nurse at the provincial hospital in Thai Binh where Sy Tuan had been taken by his father seven days earlier also contracted H5N1. Unlike Sy Tuan and his sister, nurse Thinh had not consumed duck's blood soup or been in contact with infected poultry. His only risk was that he had nursed Sy Tuan.

The Vietnamese authorities were initially cagey about nurse Thinh's illness. But Thai Binh is not the only place where apparent human-to-human transmission of bird flu has occurred. Thailand, China, Pakistan and Hong Kong have recorded similar human 'clusters' and in May 2006 Indonesia had the largest cluster of H5N1 anywhere in the world when seven members of the same family contracted the virus. To John Oxford,

Professor of Virology at Queen Mary's Medical School, London, these clusters are an alarm bell, an indication that, as in 1918, the virus is 'seeding' itself in discreet communities. Explicitly comparing the situation in Asia today to Etaples in 1918, Oxford argues that just as the crowding together of people, poultry and pigs at the British Army encampment in northern France provided the perfect conditions for the emergence of the 1918 virus so Asia's 'rice-duck-pig' farming system and the demand for poultry from fast-growing urban centres in Asia could be putting similar evolutionary pressure on H5N1.

'I view Etaples like an Asiatic family living close together, albeit a big family of 100,000,' says Oxford. 'H5 now kills roughly half the people it infects but it's not very catching. At the moment it's a slow daschund of a virus. It's when it develops into a fast greyhound that we're in for it.'[189]

But not everyone shares Oxford's pessimism. In January 2008, for instance, Bernard Vallat, director general of the World Organization for Animal Health, suggested that H5N1 was 'extremely stable' and that the risk of a pandemic had been 'overestimated.'[190] Although Vallat subsequently distanced himself from those remarks, his scepticism was echoed by Dr Paul Offit, a vaccine specialist at Children's Hospital of Philadelphia. 'H5 viruses,' he told the *New York Times*, 'have been around for 100 years and never caused a pandemic and probably never will.'[191] Jeffery Taubenberger, a molecular pathologist at the US National Institute of Allergy and Infectious Diseases who sequenced the genome of the 1918 virus in 2005, is similarly sceptical, pointing out that to cause a pandemic an avian virus would first have to become widely transmissible to humans and that 'history suggests that this may be a difficult challenge for influenza viruses.'[192] Even David Nabarro, the senior United Nations coordinator for human and avian flu, who in 2005 was one of the loudest prophets of doom, admits to being less worried about H5N1 these days, preferring to stress the threat from influenza A viruses in general as well as other emerging animal diseases.[193]

***

To understand whether H5N1 is what the writer and cultural critic Mike Davis calls 'the monster at our door' we have to return to the beginning of the present phase of the story, to China in 1997.[194] Specifically, we have to travel to Guandong, the massive province near the border with Hong Kong which Davis calls a 'postmodern Manchester' because of its hub role in China's economic miracle. Spurred by the production of trainers, sports clothes and cheap toys and electronics, Guandong's GDP grew by an astonishing 13.4 per cent between 1978 and 2002, according to Davis. But the province, which encompasses Canton and the industrial boom town of Shenzhen, is not only the epicentre of China's export-led manufacturing revolution, it is also the epicentre of bird flu. Guandong is home to 80 million people – 60 per cent of China's population. In addition, some 30 million migrant workers visit Guandong every year in search of work. To feed all those people requires millions of chickens, as well as thousands of pigs and other more exotic forms of meat – much of it imported from Laos via Vietnam. In addition, hundreds of species of wild migratory birds including geese, terns, shearwaters and gulls weather in Guandong during spring and early summer, before migrating north to Siberia and west to Africa in the autumn.

But the crucial factor may be the province's temperate climate and its centuries-old agricultural practices. For more than 300 years Guandong farmers have maintained a 'rice-duck-pig' system. To keep their rice free of insects they release flocks of ducks onto flooded paddy fields. When the rice blossoms they remove the ducks and put them on waterways and ponds. Then, once the rice has been harvested, they return them to the fields where they eat the grains of rice that have fallen to the ground.

The result are plump flocks of ducks at no cost, and neat, clean fields. Unfortunately, many farmers also raise chickens and pigs near the fields making the rice paddies the perfect laboratory for the reassortment of avian influenza viruses. Ducks are what bird flu experts call 'silent reservoirs': they carry and excrete H5N1 and other viruses without falling ill or displaying other obvious signs of infection. When chickens come into contact with diseased ducks, they pick up the virus from their excreted faecal matter. This is often the first indication that something is amiss for chickens, unlike ducks, are highly susceptible to bird flu: one moment they are clucking contentedly, the next they are staggering from side to side as their brains, stomachs, lungs and eyes leak blood in a body-wide haemorrhage. It is not for nothing that farmers describe such infections as 'fowl plague'. In addition, both chickens and ducks can also transmit bird flu to pigs and because swine can simultaneously be infected with human influenzas this makes them the perfect 'mixing bowl' for the reassortment of avian and human strains.

Scientists hypothesize that it is when these avian and human strains swop genes – reconfiguring the surface HA and NA proteins – to produce a new type of virus against which people have no antibodies that you get pandemics. Indeed, this is what appears to have triggered the 1957 'Asian flu' and the 1968 'Hong Kong' flu, caused respectively by H2N2 and H3N2 – hybrids, containing both bird flu and human influenza genes. Furthermore, sequencing studies of the 1918 virus suggest that it is the likely ancestor of all four of the human and swine H1N1 and H3N2 lineages, as well as the now

Figure 6.1 Two women selling ducks at Hanoi wet market
Credit: Mark Honigsbaum

extinct H3N2 lineage (in all, scientists have identified 16 HA and nine NA subtypes, though it is possible there are others which have yet to be classified).[195]

However, scientists also theorize that pandemics can be sparked by the process of antigenic drift alone. All that would be required would be for a bird flu virus to passage through sufficient numbers of chickens making 'copying errors' as it went along. Such mistakes would result in subtle changes to the antigenic sites – the points of the HA and NA molecules where antibodies usually bind. This is what appears to have happened in May 1997 when a three-year old boy in Hong Kong contracted H5N1 and died. It was the first time scientists had recorded the transmission of an avian influenza virus directly from birds to humans. Until 1997, such an event had been thought impossible. Quite simply, bird viruses were thought to lack the correct shaped HA to lock onto human receptor cells in the trachea and lungs. However, when scientists isolated and analysed the genome of H5N1 they got a shock – segments of the HA molecule had mutated so as to bind to both human and avian receptor sites. Just as worrying was the manner of the boy's death. He had been admitted to hospital with an ordinary upper respiratory tract infection – a sore throat and fever – but after being treated with aspirin he had developed Reye's syndrome and multi-organ failure, dying of acute respiratory distress with pneumonia 12 days later. Scientists would later discover the reason for the boy's highly pathogenic reaction to the virus – the new strain of H5N1 contained another mutation, this time on its NS gene, which induced a marked cytokine imbalance.[196]

In November, with the further deaths of a 13-year-old girl and a 54-year-old man, experts' concerns deepened. Then, in December came more cases. As international film crews descended on Hong Kong, the health department closed the city's wet markets and began an extensive cull that finally ended with the slaughter of 1.5 million chickens. But the most worrying aspect was the profile of the victims. Although the virus had attacked people of all ages and sexes, three of the six victims – half of those who died – had been young healthy adults in the prime of life. In other words, just like 1918. As Robert Webster, a virologist at St Jude's Children's Research Hospital in Memphis, Tennessee, and one of the world's leading bird flu experts, told the author Pete Davies soon afterwards, were it not for Hong Kong's extensive chicken cull, 'I would predict that you and I would not be sitting here talking now because one of us would be dead.'[197]

Between 1997 and 2002 H5N1 continued drifting. The first indication that it was evolving further pathogenicity came in December 2002 when wild waterfowl began dying in two popular Hong Kong parks. The victims included not only geese and swans but mallards, the virus's natural host. The outbreak challenged the then theory that H5N1 and other bird flu viruses were in evolutionary stasis in wild bird populations and only became pathogenic when they infected domesticated poultry. The emergence of a new strain that was also deadly to ducks suggested that the evolution of H5N1 was a two-way street. Somehow, in the passage of the virus between aquatic and terrestrial birds, the process of antigenic drift had been accelerated. Some scientists suspected that the use of unlicensed poultry vaccines in Guandong was to blame because the vaccines were putting the virus under selective pressure; others that the lethal strain might be being spread by wild ducks during their annual migration to Siberia and Alaska. Whatever, scientists concluded that the new strain, labelled genotype Z – or Gen Z for short – was here to stay and, sure enough, a few months after Hong Kong closed its parks and initiated a further cull of wild birds there were new human outbreaks.

In February 2003, Gen Z was implicated in the death of a seven-year-old girl in Fujian, China. The girl was buried before her cause of death could be established, but soon after her family's return to Hong Kong her father and brother also fell ill. The boy developed symptoms of respiratory distress but survived; however, the father died. Reagent tests showed both had been infected with the same strain of H5N1 that had been killing the birds in Hong Kong's parks.

By 2004 the highly pathogenic virus was infecting flocks throughout Southeast Asia and both Thailand and Vietnam were forced to initiate extensive culls to safeguard the reputation of their poultry industries. At the same time Gen Z continued to spark occasional human infections characterized, like the 1997 Hong Kong case, by severe respiratory symptoms. However, it was not bird flu but a completely different virus which forced scientists to look more closely at the pathology of these respiratory symptoms and prompted comparisons with the cyanosis and lung damage observed in 1918.

In early 2003 the WHO began hearing reports that Chinese officials in Guandong were battling an outbreak of an unusually powerful flu accompanied by 'atypical pneumonia.' In fact, it wasn't flu but Severe Acute Respiratory Syndrome or SARS and it's cause was a coronavirus. Normally coronaviruses provoke nothing worse than a mild cold and diarrhoea. But SARS was different. Like the 1918 flu and H5N1, it caused severe respiratory distress, pneumonia, and – in the most severe cases – death.

The natural reservoir of coronaviruses are civet cats and ferrets. Both are found in animal markets in Guandong and Hong Kong, as well as on the menus of upmarket Chinese restaurants (ironically civets are thought to provide immunity against influenza). Just as Guandong's unique disease ecology

– the pressure of so many people living in close quarters with so many animals – had led to the emergence of the first bird flu infections in 1997 so, scientists theorized, the consumption of civet cats had now prompted the coronavirus to leap the species barrier. Even more alarming, SARS, unlike H5N1, appeared to be what virologists call a 'super spreader' – at one Hong Kong hospital as many as 125 people fell ill after the introduction of the virus by a single patient; in another case the WHO traced a chain of international infections to a doctor who had checked into the Metropole Hotel in Hong Kong after visiting a patient in Guangzhou.[198]

By the end of February SARS had spread to Vietnam, Singapore and Canada and panic was widespread. In an attempt to halt the march of the virus, Toronto airport closed its runway and the WHO issued a global health alert advising travellers to avoid Hong Kong and Guandong. As in 1918, vendors of face masks and disinfectants enjoyed brisk sales and it soon became commonplace to see Hong Kong taxi-drivers spraying their cabs and advertising the fact on banners trailing from their windows. But for all that cable news channels flashed 'Pandemic Threat', SARS was not quite as super a spreader as had first been feared. Unlike the flu virus, which can be carried long distances in respiratory droplets expelled when people cough or sneeze, SARS spreads inefficiently in aerosol form. Moreover, SARS takes five days to incubate and symptoms become apparent long before it reaches peak infectiousness, making it ideally suited to old-fashioned control measures such as quarantines and isolation.

Nevertheless, the final report on SARS made sober reading. In the space of five months, the virus had spread to 26 countries and caused 8,500 infections. Although just 916 had people died worldwide – an 11 per cent mortality rate – in those aged 60 plus the mortality rate had been as high as 55 per cent.

Moreover, many of those who had suffered severe reactions appeared to have been victims of a 'cytokine storm' – the same over-reaction of the immune reaction seen in severe H5N1 infections. Such reactions tended to occur in the second week of the infection when the SARS virus moved from the upper respiratory tract deeper into the lungs triggering Acute Respiratory Distress Syndrome (ARDS). This is a scorching of lung tissue caused by the release of cytokines to the seat of the infection. Cytokines are the first line of defence against viruses, part of the body's *innate* immune response. When your body is under attack from a pathogen it has never seen before cytokines activate the B- and T-lymphocytes in white blood cells to repel the virus. The cytokines consist of pro-inflammatory cytokines, of which the most important are Tumor Necrosis Factor-alpha and InterLeukin-6. The problem is that if it is a completely novel virus – one which your body had never seen before and against which you possess no antibodies – the inflammatory response may be too strong. In essence the cytokines go into overdrive, prompting tissue to break down and fluid and blood to leak out into the lungs.

If you could place a miniature camera in the eye of the 'storm' you would see the alveoli, the tiny air sacs at the end of the bronchioles, turn pink and glassy as they became choked with fluid and debris from ruptured capillaries. Eventually, the space between the sacs would be filled with so much material that the lung tissue would take on the appearance of a single 'consolidated' mass. Indeed, scientists who have studied the passage of SARS in macaque monkeys frequently comment on this consolidation of lung tissue just as British Army pathologists did in 1918 in the case victims of gassing and severe influenza.[199] The supposition is that in all such cases the consolidation is the result of a similar overreaction of the immune system to a novel foreign invader.

Coming in the wake of the 9/11 terrorist attacks on the United States, SARS reminded the word that, in the words of Malik Peiris, the Hong Kong microbiologist who led the public health response to the virus, 'nature remains the greatest bio-terrorist threat of all.'[200] One could argue that there were other parallels too, especially in the way that Al Qaeda's murderous attacks on the World Trade Center and the Pentagon provoked the subsequent US-led invasion of Iraq – what could be seen as an over-exuberant response by the Bush Administration to the perceived attack on the American 'body politic.' But perhaps the most important lesson of SARS was how little things have changed in the last 120 years. In 1890, following the similarly sudden emergence of the Russian flu, experts pointed the finger at China and other 'silent spaces' on the map where the vastness of the territory and the seeming impossibility of getting accurate clinical and epidemiological information under-mined the possibility of mounting an effective public health response. Despite improved global systems for monitoring emerging pathogens these blank spaces remain: although the WHO funds an international network of local reporting centres, accurate and timely information out of Guandong still depends on the grace and favour of the Chinese Communist authorities. Similar political considerations inhibit the sharing of information by the Burmese military junta and the North Koreans, while in Laos, Cambodia and large parts of sub-Saharan Africa the absence of developed healthcare systems make the detection and notification of new pathogens similarly hit and miss. This is why, in spite of the WHO's success in containing SARS, experts reporting to the Royal Society in 2004 sounded a note of caution. According to Professor Roy Anderson, the Rector of Imperial College, London, and an internationally renowned epidemiologist who was then Chief Scientific Adviser to the Ministry of Defence, as much as the WHO's handling of the

SARS crisis had restored faith in the agency the world had also been 'very lucky.' It was only thanks to the low transmissibility of SARS and the fact that China and other Asian countries were able to introduce 'fairly draconian' public health measures that disaster had been averted. In north America, where people tended to be more litigious, and in western Europe, he predicted mass quarantines would have met greater resistance. The persistence of SARS in animal reservoirs meant that further outbreaks were inevitable but the real global threat came from the emergence of an antigenically novel influenza virus. 'One of the major dangers arising from the effective control of SARS is complacency,' wrote Anderson and his colleagues. 'Sentiments of the type "we have been successful once – we will be again" may be far from the truth.'[201]

***

Jeremy Farrar is far from complacent about the threat posed by H5N1 but nor does he see reasons to be overly alarmist. As director of the Oxford University Clinical Research Institute at the Hospital for Tropical Diseases in Ho Chi Minh City, Farrar has been monitoring H5N1 outbreaks in Vietnam since early 2004. In January of that year, on the eve of the Tet festivities, Farrar took a call from Professor Tran Tinh Hien, the vice-director of the hospital. A clinician and tropical disease specialist, Farrar has worked closely with Hien and the hospital since 1996 on research into infectious diseases, such as dengue, malaria, typhoid, TB and encephalitis. Hien had just admitted an eight-year-old girl with acute pneumonia and wanted Farrar to check her nasal swab for type A influenza.

The team immediately ran a test and confirmed the girl had H5N1. After taking the girl's history, Hien concluded she'd probably caught the virus from her pet duck who had died a few days earlier. Hien administered antibiotics and oxygen and prescribed a course of oseltamivir (Tamiflu) – a neuraminidase inhibitor that prevents the influenza virus from reproducing and leaving the host cell. The treatment was successful and the girl survived. However, other cases where patients arrived at hospital too late for prompt intervention had very different outcomes. In particular, Farrar recalls the case of a woman from Dong Thap, a province in southern Vietnam, who arrived at hospital with a nasty pneumonia.

'In the morning she was sitting up and talking but by evening she was moribund and in a coma,' he said. 'The following morning she was dead. When we looked at a radiograph she was missing half of her lung.'[202]

It wasn't long after that he and Hien began seeing other alarming cases, including a 16-year-old girl with acute cyanosis. Once again they were forced to intubate and place her on a mechanical lung. The girl survived a couple of days before dying of respiratory failure.

'It was a very frightening time,' says Farrar. 'At that point we didn't know very much about the virus's clinical pathology or how infectious it was. Was it the same as the Hong Kong strain of H5N1 or was it like SARs? Were other family members at risk? Could we catch it, were our families at risk?'

By the end of 2004, he and Hien were able to make a more balanced assessment. In all, Vietnam had suffered 29 cases of which 20 had resulted in death. Clearly H5N1 was not nearly as infectious as SARS. Nevertheless, the mortality rate was high and Farrar worried that the patients who made it to hospital could be just the tip of the iceberg.

'The crucial question,' he told me at the time, 'is are we only seeing a small number of cases because we are only looking at patients who present with the most severe symptoms or is the virus actually much more prevalent than that? In other words is this just a very nasty infection in a few people, or a wide-ranging infection, one that has the potential to become a global pandemic?'

Four years later Farrar is in a better position to answer that question. At the time of writing there have been 106 human cases in Vietnam and 52 deaths. Elsewhere in Southeast Asia it's a similar picture: Indonesia, which now tops the WHO's H5N1 league table, has suffered 135 infections since 2005 and 110 deaths. The big unknown is China – officially the Chinese have notified the WHO of just 30 cases. Then there are the mounting concerns about H5N1 outbreaks in poultry in India, Pakistan and Bangladesh. The most densely populated country in the world, Bangladesh produces an astonishing 220 million chickens and 37 million ducks every year. About four million Bangladeshis are directly or indirectly associated with poultry farming. That presents the virus with plenty of opportunities for further mutation and even reassortment.

But while for many Bangladesh is the stuff of nightmares, Farrar argues that Armageddon may never come. 'We've been having major global poultry outbreaks since 1995 but so far there's no evidence of widespread human infections or efficient transmission between people,' he says. 'The fact that there have been fewer than 400 human cases in 13 years shows that there must be very severe biological constraints on the virus that may not allow it to adapt fully to humans.'[203]

One reason H5N1 may not have triggered more infections is that it binds to cells deep in the lungs. By contrast, human flu viruses carry adaptations allowing them to bind to cells in the nasal passages and upper respiratory tract. This suggests that H5N1 is not that easy to catch, at least in aerosol form, and that it takes repeated exposure – as in the case of a parent nursing a sick child – for a person to contract the virus.

A second reason for optimism is that more than 90 per cent of H5N1 case clusters have occurred in blood-related family members. In other words, the virus tends to be passed between parents and their children, or brothers and sisters, but is rarely transmitted between husbands and wives. 'This suggests there may be a rare genetic trait that makes some people more susceptible to the virus,' says Farrar.

Recent serological studies in Vietnam and China have also found very low levels of antibodies to H5N1 in the general population. While for those unfortunate enough to be exposed to the virus and get sick that is a bad thing, from a global public health perspective it is a good thing because it means that the chances of any one individual being infected with both H5N1 and a seasonal influenza virus at the same time – the prerequisite for a reassortment event – is very low.

'At the moment, if someone gets H5N1 it's a very nasty infection, which kills a high percentage of people it infects, but there are not thousands of people getting it and being asymptomatic,' says Farrar. 'That's a very important piece of information because it means that human H5N1 infections are still an incredibly rare event.'

Finally, many scientists now argue that there is no reason to believe that just because pandemics have occurred with regularity in the past they will continue to do so with the same regularity in the future. One factor that might disrupt the 'natural' flu cycle is changing population immunity caused by increasing use of immunologically complex vaccines; another is the reassortment of pandemic strains with closely related HAs. Then there is the observation that in 1976 a fatal H1N1 'swine flu' triggered considerable alarm without triggering a pandemic and that a year later a descendant of the 1918 virus suddenly reemerged (most likely as the result of an accidental laboratory release) to reestablish postpandemic circulation with one of its own descendants, the H3N2 virus. Such occurrences have led some distinguished flu researchers to conclude that 'there is no predictable periodicity or pattern' of major influenza pandemics and that 'all differ from one another.'[204]

**

One thing that might help us to assess the chances of H5N1 or some other variety of bird flu triggering a pandemic is a better understanding of where the 1918 virus came from and how it acquired the mutations that transformed it from a virus that merely sickened chickens to one that killed millions of people around the globe. Fortunately, in recent years modern genomics has at last brought this holy grail of influenza research within our grasp.

A former student of Sir Charles Stuart-Harris – a member of the team that isolated the first influenza virus in London in 1933 – John Oxford has spent most of his professional career trying to unravel the puzzle of the 1918 flu. Like his colleagues he has looked for clues to its virulence in analyses of antibody titers of 1918 flu survivors from the late 1930s and in strains of H1N1 – ancestors of the original 1918 virus – that continue to infect swine to this day. However it wasn't until the early 1990s when a colleague informed Oxford that there were hundreds of pathology specimens – slivers of lung tissue – taken from British victims of the 1918 flu in the bowels of the Royal London Hospital in Whitechapel, east London, near where Oxford works that he had his eureka moment. What if he commissioned his students to search those specimens and they discovered intact lung tissue containing the 1918 virus? Using the latest molecular pathology techniques, he should be able to extract sufficient viral genetic material to sequence the 1918 virus's RNA and determine whether, like H5N1, it was of avian origin. If Oxford was really lucky and got the whole sequence, then he might even be able to answer the question for which they award the Nobel prize: namely what was the crucial mutation which turned the 1918 virus into a super-spreader?

Unbeknown to Oxford, on the opposite side of the Atlantic another scientist was about to have a similar epiphany. Jeffery Taubenberger is both a biologist and a qualified medic, a rare combination. After a residency at the National Cancer Institute and a stint at the National Institutes of Health, in 1993 Taubenberger joined the Armed Forces Institute of Pathology (AFIP) where he set up a state-of-the-art molecular pathology lab. A five-storey concrete bunker set in the rolling hills of Bethesda, Maryland, the AFIP is an unlikely place to attempt cutting edge research. But with the help of a colleague, Anne Reid, Taubenberger developed a technique for extracting fragile viral genetic material from damaged or decayed tissue. Using this technique, in 1995 Taubenberger and Reid were able to demonstrate that bottle-nosed dolphins which had recently washed up on the coast of New Jersey, and which most scientists believed had been killed by red tide toxins, were actually the victims of a rare morbillivirus.

Soon after Taubenberger read how researchers had used polymerase chain reaction (PCR) – a technique which allows scientists to take tiny fragments of genetic material and make multiple copies of its RNA – to isolate the genes for colour blindness from the preserved eyeballs of a famous chemist, John Dalton. Taubenberger wondered whether he might be able to solve other medical puzzles using PCR and the extraction techniques he and Reid had pioneered at the AFIP. That's when someone suggested the 1918 flu. Buried in the AFIP's archives were three million pathology specimens, including some 100 autopsy samples taken from 1918 flu victims. Like

Oxford, Taubenberger wondered whether these slivers of lung tissue preserved in formaldehyde and embedded in paraffin might contain tiny fragments of the 1918 virus.

After an agonizing year of negative results, Taubenberger got his answer. In 1996 he isolated the first influenza-positive sample from a soldier who had died in September 1918 at Fort Jackson, South Carolina. The following year, he found a second sample in the lung tissue of a soldier who had died in the same month at Camp Upton, New York. But while Taubenberger was able to sequence five of the eight genes of the 1918 virus from these samples, he did not have sufficient material to complete the analysis. Nevertheless, in March 1997 Taubenberger published his findings in *Science*.[205] The announcement catapulted him onto talk shows and the front pages of newspapers worldwide. It also alerted Oxford that he had a competitor.

To his credit, Oxford quickly realized that Taubenberger was better equipped to extract RNA from the lung tissue of the British victims of the 1918 flu that his students had by now unearthed from the stores at the Royal London Hospital, so he forwarded his samples to Taubenberger in Bethesda. The sequences isolated from the British soldiers proved an almost exact match for those isolated from the American soldiers. But the key would be a young Inuit woman who'd died of flu at Brevig, Alaska, in 1918. 'Lucy,' as the woman was dubbed, had been buried in a mass grave which had then frozen in the permafrost. With the permission of local people Lucy's body was exhumed and a section of her frozen lung tissue extracted and forwarded to Taubenberger. Those samples enabled Taubenberger and Reid to complete the sequence of the 1918 virus. Using a technique called plasmid-based reverse genetics, other researchers at the US Centers for Disease Control and Prevention (CDC) in Atlanta, Georgia, then made new copies of the virus and infected mice to test the virus's virulence. Their findings, announced in *Nature* in October 2005, came as a shock to sceptics who considered that for a bird flu virus to become infectious in humans it would first have to undergo reassortment with a mammalian strain. None of the eight genes came from a strain that had previously infected humans. Furthermore of the 41 amino acids that have been shown to be targets of the immune system and subject to antigenic drift pressure in humans, 37 matched avian sequences. This suggested that the virus began as a bird-adapted strain that with just a handful of mutations had become infectious to people. 'It is the most bird-like of all mammalian flu viruses,' Taubenberger told *Nature*.[206]

There was another shock too. When researchers infected mice with the virus it generated 39,000 times more virus particles than a modern flu strain. Terence Tumpey, the lead CDC researcher who conducted the trials in a biolevel 3 safety containment lab, discovered that this virulence depended on a complementary relationship between three genes. It was a 'combination effect,' he told *Nature*.[207] Somehow, the three genes worked synergistically to produce a triple whammy.

Further evidence for the sudden emergence of the virus from an avian source came from regression analyses of ancestral strains of the 1918 virus. Taking flu viruses isolated between 1930 and 1993, Taubenberger and his colleagues calculated the number of amino acid changes in the surface proteins required for divergence from the 1918 virus and then plotted those divergences back in the time. The lines crossed in the period 1915–17, suggesting that if the virus had emerged directly from an avian source, the event must have happened very suddenly, shortly before the virus became infectious to people.[208] But Taubenberger also noticed an oddity: the chain of nucleotides that coded for the amino acids governing the shape of the HA surface protein in the 1918 virus were arranged differently from those of other known bird flus. That could mean one of two things: either bird flus had evolved over the decades and in 1918 the amino acids in bird viruses were coded differently, or the 1918 virus came from a bird no one had considered before.

In an attempt to answer that question Taubenberger approached the Smithsonian Institute. At the Natural History Museum in Washington the Smithsonian keeps thousands of preserved bird specimens from the early 20th century. Zeroing in on waterfowl collected around 1918, Taubenberger took tissue from the birds' excretory tracts and examined it for flu viruses. Taubenberger obtained positive results in six cases. However, the genetic coding was exactly like that found in contemporary bird flu viruses. In essence, Taubenberger was back to square one: either the virus had come from some unknown avian source – a strain that was evolutionarily isolated from the typical wild waterfowl influenza gene pool – or it had come from some other as yet unidentified host.

However, that was not an end to the conundrums. If Taubenberger's analyses of the rates of mutation of the 1918 genome were correct and the virulent strain had emerged as early as 1915–17 why hadn't it triggered a major disease event sooner? After all, there was plenty of evidence from both the US and the UK for the high prevalence of influenza and pneumonia in the two years prior to 1918. Furthermore, why, if people were already manufacturing antibodies to the virus in those years, was not a greater section of the population immune or at least partially protected when the pandemic did occur?

Unfortunately, to answer these questions Taubenberger would need more virus samples – in particular he would need to know what flu viruses were circulating in 1915–17. He would also need serum samples. And, ideally, he would need virus samples both from the spring wave in 1918, which lacked the virulence of the autumn wave, and the later winter 1919 wave (three waves in less than a year is very unusual and suggests that either the surface proteins of the 1918 virus drifted more rapidly than other pandemic strains or the virus had an unusually effective way of evading the human immune system). Armed with such samples Taubenberger might also be able to shed light on the unusual virulence observed in young adults. Influenza and pneumonia death rates for 15–34 year olds were more than 20 times higher in 1918 than in previous years. Based on serological and epidemiological data, researchers hypothesize that an H1 influenza virus circulated until 1889 when it was replaced by the Russian flu – an H3 virus. The H3 subtype then prevailed until 1918 when it was replaced by the new pandemic strain of H1. If this theory is correct then only people born before 1889 – those aged 30 and over in 1918 – would have possessed any H1 antibodies, while those born after 1889 would have been immunologically naïve, hence the high mortality observed in the median age groups.[209] But, again, without blood serum from this period that can be tested for antibodies, Taubenberger says it is 'exceedingly difficult' to say what role host factors may have played in virulence.[210]

Nor can Taubenberger say whether the virus emerged first in the United States, northern France, or somewhere else. 'I am literally agnostic about the origin of the 1918 virus in terms of geography,' he says. 'The actual data do not allow me to even pick a hemisphere for the origin of the virus. The historical data are all extremely limited – no cases, no isolates, no serology – so I do not believe it is currently possible to place the origin of the pandemic.'[211]

These limitations, he points out, also apply to Oxford's Etaples theory. Without viral samples from men stationed at the camp it's impossible to say whether the virus originated there. In any case, Taubenberger believes the pathology of purulent bronchitis is 'not so supportive of flu.'

As to timing, Taubenberger says that based on his regression analyses it is 'possible' the 1918 virus was circulating in human populations a couple of years before the pandemic. However, he thinks it more likely that circulation began in the 1917–18 flu season with H1N1 gradually replacing the previously circulating strain until the sudden outbreaks observed in March 1918. 'By March,' he concludes, 'it was already in the US, Europe, and possibly China.'

Predictably, Oxford is not happy with such equivocation. 'It's got to start somewhere, there's got to be a first case,' he says. The notion that the outbreak could have originated in Kansas he calls the 'yellow brick road theory.'[212] Although in principle, migratory birds can carry avian flu viruses to any part of the globe, evidence from the current H5N1 outbreak suggests that the Eurasian and North American pathways are quite distinct.[213] Oxford also points out that so far no one has produced any evidence to show soldiers kept ducks and chickens at Camp Funston. By contrast, he says: 'We know that all the circumstances involved in the emergence of these viruses were at Etaples. It's on this huge migration route on the Somme [and] there are all these pictures of soldiers with birds, and there are piggeries.'[214]

However, Taubenberger points out that although his research is consistent with the possibility that the precursor to the 1918 virus was hidden in an obscure ecological niche before emerging in humans, no large die-offs were observed in wild waterfowl or domestic poultry at the time, leaving it unclear what the source of the 1918 virus was and how the virus adapted to humans.

The difference of opinion is not likely to be resolved any time soon. The origins of the 1918 pandemic are a puzzle within a riddle within an enigma. Taubenberger's sequencing of the virus's genome both supports the theory that it was a uniquely lethal strain and also that with a little serendipity – the mutation of just a few amino acid sequences here and there – it could happen again. 'Even with the whole 1918 genome available, we may never fully understand what happened in 1918,' he admits.

Even if history does repeat itself – and a novel virus suddenly emerges from a hidden avian reservoir – the impact need not be catastrophic. In April 2008, for instance, the *Lancet* published details of the best documented case of human-to-human transmission of H5N1 to date.[215] The report was the result of a rare collaboration between a Chinese medical team and the US Centers for Disease Control – in itself an encouraging development – and described how in November 2007 a 24-year-old salesman from Jiangsu province had been admitted to hospital with fever, chills and headache. Unfortunately, doctors diagnosed his illness as bacterial pneumonia rather than bird flu and five days after his admission he died of ARDS and multi-organ failure. However, staff were rather more successful in treating his 52-year-old father who almost certainly caught the strain of H5N1 from his son (the father had not been exposed to diseased poultry but had shared several meals with his son and had helped nurse him in hospital). Initially, he was given oseltamivir and other anti-viral drugs but when his respiratory status worsened doctors decided to try a novel therapy: they decided to immunize him with blood plasma from a woman who had received an inactivated H5N1 vaccine. It was the same principle as Pasteur had used when he induced an antibody response to rabies using passive serum from rabbits, and not that dissimilar to the immunization trials conducted by bacteriologists in 1918. That night the man's fever resolved itself and within three weeks he was well enough to be discharged. Although there were no controls, his recovery from what appears to have been a very nasty case of bird flu suggests that in the event of a pandemic treatment with rough and ready serum vaccines might prove just as effective as the real thing.[216]

The report also contained two further pieces of good news. Despite having had close contact with 91 other people, the father and son were the only two people to fall ill and from serological tests none of their contacts appeared to have been infected with the virus. Nor was there any evidence that the strain isolated from the men had mutated so as to spread more efficiently.

'There is no doubt a pandemic could be caused by H5N1 or any other avian influenza virus,' concludes Farrar. 'Having said that, the chance of a purely avian virus crossing the species barrier and causing a global pandemic appears to have been a very rare and possibly unique event in human history. Learning what prevents this jump between species is what may save us from another.'

7

Britain, Summer 2012

London, 20 July – *Several members of the Chinese table tennis team were admitted to an isolation ward at the Olympic Village in east London last night following reports that a deadly new form of influenza has broken out among athletes attending the 30th International Olympiad.*

The decision to quarantine the athletes, who include the Chinese women's team coach Deng Yaping, was taken by senior Department of Health officials following consultation with the World Health Organisation's rapid influenza response team and the International Olympic Committee late yesterday afternoon.

It is understood that Yaping, a five-times Olympic gold medallist, is in 'critical condition' with pneumonia and is receiving oxygen via a ventilator. Three other members of the Chinese women's table tennis team with whom she had been sharing an apartment in the Olympic Village are also reportedly running high temperatures and having difficulty breathing.

Professor Lindsey Davies, national director of pandemic influenza preparedness, said the athletes were being quarantined at the village's medical centre as a 'precaution' and there was presently no cause for alarm. However, because of the risk of human-to-human transmission the WHO had announced a 'phase four' international alert and the UK had raised its alert status to level two.

'If the athletes' condition worsens we may have to move them to a better-equipped hospital but for the time being we think it

better to try to treat them on site,' said Professor Davies. She added that at present there were no plans to extend the quarantine to the rest of the Olympic site.

It is understood that immediately prior to arriving at the Olympic Village Yaping had been visiting her family in Zhengzhou, the capital of Henan province, an area of China where bird flu is endemic. Blood samples from Yaping and the other athletes were forwarded to the National Institute for Medical Research (NIMR) in Mill Hill for analysis last night. However, medical sources said that initial reagent tests conducted by the WHO's rapid response team suggested the virus was not H5N1 avian influenza but an 'H2' – a type of human influenza virus that last triggered a pandemic in 1957.

Dr Alan Hay, the director of the World Influenza Centre at the NIMR, was not available for comment. However, Professor John Oxford, a leading virologist at Queen Mary's Medical College, London, said that if it was an H2 that could be 'very bad news indeed' as the virus stopped circulating in 1968 meaning that only those aged 42 and over would now possess antibodies to it.

'It's possible we've all had our eyes on the wrong target,' said Oxford. 'We know that H2 viruses, unlike H5N1 and other avian flu viruses, are capable of rapid human-to-human transmission. My fear is that in a confined community like the Olympic Village the virus could be amplified many times over.'

With 17,000 athletes from more than 200 countries due to arrive at the Olympic Village in Stratford this weekend ready for the opening ceremony in seven days time the outbreak could not have come at a worse time. London Resilience, the capital's emergency response agency, plans to visit the site today to disinfect public areas and check hygiene standards in the restaurants and bars in the village's 'International Zone.' Sebastian Coe, the chairman of the London Organizing Committee, said officials would also be distributing hand dispensers and disposable paper towels to athletes and their families.

'There is absolutely no need for panic,' he said. 'The games will go on.'

The Department of Health said that at present there were no plans for travel restrictions to and from the site. Instead, it would be distributing leaflets to households in Stratford emphasizing personal hygiene measures, such as regular hand washing, and advising people worried they might have flu to call the National Flu Hotline.

'We will be issuing further information as soon as it becomes available but for the moment it's business as usual,' said Professor Davies.

The above scenario is hypothetical, a projection of how the next influenza pandemic might start. Perhaps the pandemic will come in 2012, or perhaps it will happen sooner. It may start in China, as several experts have been predicting for some years now, or perhaps the seat of the outbreak will be Bangladesh. Perhaps the infection won't be introduced to the Olympic Village by a Chinese athlete but by a Bengali chef working in one of the concessions in the International Zone – or perhaps, it will arrive in London's Docklands via a banker or journalist returning to Canary Wharf from an overseas assignment. Whatever the case, we cannot simply wish the next pandemic away. As Britain's Chief Medical Officer Sir Liam Donaldson told the BBC in an interview in 2005 a new pandemic is a question of 'when, not if.' 'We can't make this pandemic go away, because it's a natural phenomenon, it will come,' he said. 'But what we can do is limit its impact.'[217]

Sir Liam's frankness provides a stark contrast with the reticence of his predecessor. In 1918, when contemplating a similar pandemic threat, Sir Arthur Newsholme opted to do very little, leaving it to local medical officers to decide what advice to give the public and how to go about treating people and disposing of the dead. Today, hardly a week passes without some announcement from the Department of Health or the Cabinet Office's Civil Contingencies Secretariat (CCS) updating the government's 'pandemic preparedness' plans. Thus we know that the government's contingency plans include the stockpiling of 14 million doses of oseltamivir– enough to cover 25 per cent of the British population – and 'sleeping contracts' for the fast-track manufacture of 120 million doses of vaccine so that everyone in Britain can receive at least two shots. As soon as the WHO notifies world governments of clustering with limited human-to-human transmission – the trigger for a phase four alert – the government will launch a major advertising campaign advising that pandemic influenza could arrive in Britain within two to four weeks and explaining how people should stockpile supplies of food, water and vital medicines. Then, once the virus reaches Britain and begins causing widespread morbidity and death (UK alert level 4), the Chief Medical Officer will go on TV to address the nation. This will be a critical moment. Pandemic planners have identified communications as the key to the government's successful management of the pandemic: if the CMO's tone is too shrill he risks panicking the nation; if he is too relaxed people may not take the threat seriously enough. The message he has rehearsed is that those who are ill or fear they have the symptoms of flu should stay at home to reduce the risk of spreading the virus, but there is no reason why everyone else shouldn't travel to work as normal. The key phrase will be 'business as usual.' The government's biggest fear is not so much the initial wave of morbidity as the knock-on effects on the British economy. Pandemic models suggest that if people can be made to feel safe about travelling to work then absenteeism can be kept to around 25 per cent. But if people panic and stay at home absenteeism could reach 50 per cent severely impacting everything from food deliveries to telecommunications to the ability of utility companies to continue generating electricity. 'There are lots of unknowns today,' admits Professor Davies. 'We are very conscious of the fact that the British economy is far more reliant on international trade than it was in 1918. Everything is much more interconnected and interdependent.'[218]

A softly spoken woman with short blonde hair and glasses, Davies got the call to become national director of pandemic planning in 2006. A qualified doctor, Davies had previously held senior health posts in Derby and Nottingham and been regional director for public health for the Midlands. But though Davies was used to planning for disease disasters such as SARs, until 2006 she'd never considered how the National Health Service might respond to a re-run of 1918. Today, having spent two years talking to experts about everything from the provision of face masks to equipping hospitals with sufficient ventilators and oxygen, both she and the government are much better prepared. Nevertheless, Davies is the first to admit that in the event of a pandemic – especially one with an attack rate higher than 40 per cent – the NHS could be rapidly overwhelmed.

'There's no point in pretending that we'll be able to cope in the normal way. There will be a lot of people being cared for in the community with severe symptoms that doctors are not used to dealing with.'

The problem is although Britain has many more doctors and nurses than it did in 1918 and more intensive care units, it also has a far larger population (approximately 60 million versus 34m in 1918) with far higher expectations about the quality of medical care it is entitled to receive. If just a quarter of the British population were to fall ill – the same number who were affected in 1918 – some 15 million people might seek advice and/or treatment. In theory, says the Royal College of General Practitioners, consultations would jump from an average of 30 per week per

100,000 patients to between 5,000 to 10,000 a week, overwhelming the capacity of the typical Primary Care Trust. As the
college's chairman Nigel Mathers told a house of lords select
committee looking into the government's pandemic planning
in 2005, 'Our surge capacity would not be able to cope
with demand like that and we would need some alternative
provision.'[219]

The impact on hospitals is even more critical. Should ten per
cent of the affected population develop pneumonia or an
ARDS-like illness then potentially 1.5 million people might be
in need of intensive hospital care and ventilators. But at
present England has just 110,000 acute hospital beds and a
further 1,800 intensive care beds. Of course, not every part of
the country or every hospital would be affected equally
severely or at the same time – as in 1918, the epidemic waves
might peak sooner in London than in Manchester.

In 2005 the Intensive Care Society used 'Flusurge' software to
calculate the impact a pandemic might have on the demand for
hospital beds. The calculations make sobering reading. In an
eight-week epidemic with a 25 per cent attack rate, the same as
in 1918, it calculated admissions to hospitals would be running
at approximately 5,500 a day or 35,700 a week. At this level of
admissions, demand for critical care beds would be 208 per cent
higher than the current capacity. As David Menon, Professor of
Anaesthesia at Addenbrooke's Hospital, Cambridge, and one of
the authors of the study told Jo Revill, then Health Editor of the
Observer: 'To deal with that you would have to double our
capacity, which would be impossible.'[220]

Davies is rather more optimistic about the NHS's ability to cope
with the elevated demand. One possibility being considered by
the department of health is to make oxygen ventilation available
in other parts of hospitals. Nevertheless, Davies admits that in a
very short time intensive care facilities would reach breaking

point. The challenge then would be: 'Who are you going to discharge? Are you going to give people three days and out? Those are the very real decisions people are going to be faced with.'[221]

But it is not only doctors and hospitals that would struggle to cope. If 1918 is any guide, chemists would also come under intense pressure. At the top of people's shopping lists will be analgesics, antibiotics and antivirals. However, both oseltamivir and zanamivir (Relenza) – another antiviral which like oseltamivir targets the proteins on the surface of the virus – will be strictly rationed. Courses will not be given prophylactically. Instead, prescriptions will be reserved for those who exhibit symptoms of flu in the hope that early treatment can reduce the risk of severe complications. Priority will be given to children, the elderly and key health personnel such as doctors, nurses and ambulance workers. Health Protection Agency personnel and other professionals involved in the front line of the emergency response, such as policemen, firemen and coroner's staff, would also get priority but, significantly, not their families. For some this could prove a problem.

'The current pandemic preparedness plan puts me in a dilemma,' explains Richard Jarvis, a Merseyside health protection consultant and a member of the British Medical Association's public health committee. 'Do I continue working and try to manipulate the system to ensure that my family get protection, or do I take myself out the frontline so that I do not put myself at risk of exposure and of infecting my family?'[222]

The provision of sufficient supplies of antivirals is critical to the success of what Davies calls the government's strategy of 'defence in depth.' If the department of health can build up a larger stock of antivirals then it will be in a much better position to reduce the incidence of severe complications and, hence, the pressure on intensive care beds. And if it can build up a stockpile sufficient to cover half the population – the current planning aim – then it would have the flexibility to prescribe antivirals to every member of a household once one person in that household falls ill. In theory this would result in a double benefit: early treatment would enable people to manufacture antibodies to the virus without becoming ill and at the same time the prescription of antivirals would help reduce people's anxieties about the pandemic.

However, it is the harder-to-predict secondary consequences of a pandemic that gives Davies sleepless nights. Britain today is a very different country from 1918. In 1918 people did not expect handouts from the state, nor did they receive them – instead they relied on extended family networks, as well as churches, local community groups and charitable organizations. At the same time, there was a strong voluntarist ethic running through society – women volunteered in their thousands for the Red Cross of joined land brigades to help farmers get in the harvest. Nor in 1918 did Britons depend on supermarkets and complex international supply chains for basic foodstuffs such as grains, vegetables and fruit. With the exception of wheat from America, the main source of carbohydrates and vegetables were local farms and allotments, and while meat and fish were often in short supply these too came from local sources. Today, however, supermarkets operate on the 'just-in-time' delivery principle meaning they only keep sufficient supplies on the shelves for three days. After that, the stocks have to be replenished. But what if international cargo flights are grounded, or the lorry drivers supermarkets rely on to deliver fresh produce to their warehouses decide to stay at home? Moreover, what if there are also high rates of absenteeism among other key workers, such as power plant operators, bank and telecommunications staff, or the police and fire services, as occurred in 1918? As supermarkets run out of food and water, ATMs run out of cash and fuel pumps run dry, will people be content to stay at home and call the National Flu Hotline for advice – as the government's current plan envisages – or will they panic? And what if people can't get access to the hotline because the operators are all busy dealing with other calls or ill themselves and bodies start to pile up? Will people be content to remain at home meekly listening to government pronouncements or will they take the law into their own hands?

A breakdown in law and order would have been unthinkable in 1918. Yes, there were sporadic mutinies at army camps – including in 1917 at Etaples – and there was also the threat of strikes by shipyard workers in Clydeside and the police force in London, but these challenges to the state's authority were driven by particular political and economic grievances and were soon assuaged. On the major issue confronting the nation, namely the war, British society was united. Besides, by 1918 Britons had grown used to the wartime privations and expected little from the state and even less from the medical profession. Today, society is far less resilient and far more individualistic, making a breakdown in law or order not only possible but probable. Indeed, you could argue that this is the real 'monster at the door.' In the event of a repeat of 1918, the department of health calculates that anywhere from ten per cent to half the UK population could fall ill over a fifteen-week period. If the virus had the same attack rate and mortality rate as 1918 (25 per cent and 2.5 per cent respectively), the deaths would number 375,000. If the attack rate was double and the mortality rate a lower 1.5 per cent, then the predicted death toll would be even higher – 450,000. But the Armageddon scenario is a 50 per cent attack rate and a 2.5 per cent mortality rate, an equation which produces a staggering 750,000 deaths.

A key issue will be disposal of the dead. In 1918 bodies were left in homes for up to two weeks as local authorities struggled to find wood for coffins and people to dig graves. The pressure today would be possibly even more intense. In the winter of 2007 the Cabinet Office's Civil Contingencies Committee (CCC) conducted a simulation exercise codenamed *Winter Willow* to test the country's preparedness for a pandemic.[223] The exercise involved some 5,000 emergency response staff across the UK – everyone from ambulance drivers, to coroner's officers to local health visitors. The full findings have yet to be published but ambulance drivers experienced a particularly high workload and struggled to answer call-outs. Another issue identified by the CCC was the need to show 'respect and dignity' to people of different faiths. In 1918 Britain was relatively homogenous. Today, Britons live in a multicultural, multifaith society. In the event of a pandemic, the voices of some faiths might prove louder than others, particularly where their religion dictates that the deceased should be buried within three days. But in the event of a pandemic giving priority to one group over another may result in accusations of favouritism, particularly as the bodies begin to pile up and the demand for burials sites becomes more urgent.

The authorities will also have to find a hygienic way to store bodies until death certificates can be issued, a requirement under British law before burial can proceed (current plans envisage the distribution of body bags and the erection of temporary mortuaries). One proposal is to bring doctors out of retirement to help issue certificates. The government is also considering suspending the requirement for coroners to hold inquests and increasing the capacity of crematoria. However, at present cremation is an option taken up by just 30 per cent of Britons and if the majority of families insist on burials mortuary attendants and grave-diggers may find themselves unable to cope. Davies is reluctant to talk about the different options under consideration but admits that it may be necessary to bury people in temporary graves, especially given the likelihood that members of the same family, not of all of whom will die at the same time, will want to be buried together.

'You would put people in individual graves and marked, but not necessarily the one they would end up in. It's not nice talking about but that's what you'd have to do.'

Another word that does not appear anywhere in the government's plans is 'curfew.' Nor will you find any references to police action or the possibility of 'martial law.' However, during the lords select committee inquiry in 2005, a senior Metropolitan Police Authority planner revealed that police had rehearsed the possibility of stationing officers at the doors of surgeries and clinics in the event that people became frustrated at rationing. And in its current plan the department of health acknowledges that, in a worst case scenario, the need to prioritize medicines 'could result in disturbances or threaten a breakdown in public order.'[224]

'There is a risk of breakdown if everyone panics and they think they must have their antiviral now,' Davies acknowledges. 'Communications will play a key part in getting people to understand what the stockpiles are for and who will get them. We need to get people on side.'

But for all that it may be rational to rehearse for *28 Days Later*-style scenarios, there is also another side to the ledger, namely that the medical profession is far better equipped to deal with influenza and the complications of viral infections of the respiratory tract that it was in 1918. Studies suggest that if antivirals are administered promptly (usually within one to two days of the onset of symptoms) they can reduce both the length and severity of symptoms. Although for obvious reasons there is no data on the efficacy of antivirals in a pandemic situation, the department of health estimates that they could lower attack rates by up to a third. This would not only reduce the workload on GPs and hospitals but, just as importantly, it could make a critical difference to levels of absenteeism. Another poss-ibility under consideration is immunization with 'mock-up' vaccines. Here the idea is to use vaccines from non-pandemic virus strains to trick a person's immune system into producing antibodies until such time as specific vaccines against the pandemic strain become available – a process which typically takes four to six months from the identification of the pandemic virus. However, many scientists have reservations about 'mock-up' vaccines arguing that they are expensive and there is no guarantee they will provide 'herd' immunity sufficient to protect the population as a whole. Having said that, a specific vaccine, once it is ready, should greatly reduce the morbidity and mortality of the follow-on waves.[225]

The biggest reason for optimism, however, is antibiotics. A major, if not the major, cause of excess mortality in 1918 were not the initial infections from the virus but the follow-on cases of severe bronchopneumonia caused by secondary bacteria. Although data from military and civilian hospitals produce different percentages for the proportion of 'cyanosis' or ARDS cases – what today we would characterize as primary viraemias – and cases of severe bronchopneumonia triggered by secondary bacteria, it seems reasonable to conclude that had doctors been able to administer antibiotics many patients would have sur-vived. Indeed, the authors of one recent analysis looking at the potential global mortality from a 1918-style pandemic conclude: 'In middle-income and low-income settings, prompt access to antibiotics could be the most affordable strategy that has the largest effect on mortality.'[226] The authors of another study point to other advantages, such as the existence of hospital-based intensive care and ventilatory support and the WHO's global influenza surveillance system which ought to give two to three weeks forewarning of a pandemic (a capability that did not exist in 1918).[227] Indeed, during *Winter Willow* assumptions about a warning period of several weeks were built into the exercise.

But what if there is no warning and the virus arrives at the Olympic Village in east London unheralded? At present Britain, like other countries, assumes a pandemic will start outside its borders, most likely as the result of a tourist or businessman importing the virus from abroad. But as the recent progress of H5N1 west across Europe from the Urals demonstrates, bird flu could just as easily be introduced directly to the UK by migratory birds. Indeed, the infection of poultry flocks in Turkey and Romania in 2006 by wild geese returning to Africa from their winter breeding groups in Siberia and China alerted European veterinary officials to just how susceptible all countries are to sudden outbreaks from animal reservoirs. Another way the virus may gain entry is via commercial poultry movements. To date just four birds in Britain have died of H5N1 as a result of natural migrations – in April 2006 when a whooper swan washed up dead at Cellardyke harbour in Scotland and in January 2008 when the carcasses of three wild mute swans were discovered at the Abbotsbury Swannery in Dorset. By contrast, there have been two costly introductions of bird flu by poultry processing companies: in February 2007 when 160,000 birds at a Bernard Mathews turkey farm in Suffolk had to be destroyed following the accidental importation of H5N1 from one of the company's Hungarian subsidiaries, and in November 2007 when the bird flu virus infected flocks at a processing facility on the Norfolk-Suffolk border owned by Redgrave Poultry (to contain the outbreak Defra erected a two-mile protection zone around the farm and, as a precaution, culled 28,000 free-range turkeys at four nearby farms). Finally, there was the case of the parrot from Surinam that died in an Essex quarantine facility in Essex in October 2005, apparently after contracting H5N1 from a consignment of finches from Taiwan.

Moreover, what if state-of-the-art medical interventions fail or do not deliver the expected reductions in morbidity and absenteeism that planners are predicting? Once there is a confirmed case of human-to-human transmission in Britain, Davies says there would be little point in quarantine measures such as closing borders or banning international air traffic as studies show that even a 99.9 per cent restriction in travel to the UK would only delay the onset of a British epidemic wave by two months. Besides, the disruption to trade would make quarantines economically unacceptable.

Internal travel restrictions, such as discouraging people from taking trains or boarding buses, are also unlikely to be effective as studies show that even a 60 per cent reduction in travel results in only a small flattening of the profile of a pandemic. Nor is the government proposing blanket bans on public gatherings, such as sporting fixtures, or universal school closures. As in 1918, decisions about whether or not to close schools will be left largely to local education authorities. This is not because the department of health thinks such closures would be ineffective – on the contrary a recent study found that closing schools could reduce the cumulative number of flu cases in the population by up to 17 per cent (this is because some children are 'super-spreaders,' shedding the virus in large quantities and infecting adults while showing few or no symptoms).[228] Interestingly, the main reason is that the government fears that if schools were to close prematurely many mothers, including working mothers, would have no choice but to stay at home to care for their children, resulting in widespread disruption to industry. Instead, the current advice calls for local authorities to plan 'both for continuing to operate *and* for possible closures [my italics].'[229]

Isn't this a bit of a fudge? Not at all, replies Davies. 'If the flu hasn't got to you do you have to close schools? No. But if it's in your community and people are very ill then, yes, it would be sensible. What we wouldn't have is a blanket policy.'

Flexibility on school closures isn't the only measure with echoes of 1918. Whether intentional or not, 'business as usual' sounds very much like Newsholme's invocation to 'carry on.' Indeed, reading the government's recommendations on 'social distancing' and personal hygiene measures you cannot help but be struck by how little has changed. In 1918 James Niven advised Mancunians that in the event they became ill they should isolate themselves in a separate room and 'in this way the spread of the disease will be delayed.' In 2008 the department of health writes that the sick should stay at home in order to protect others and 'reduce the spread of infection.' In 1918 Newsholme advised people to avoid coughing and sneezing in public and if they found it necessary to use a hand-kerchief to boil or burn it afterwards. In 2008, the department advises people to cover their noses and mouths with a 'tissue' before disposing of the said tissue by 'bagging and binning.' The department also stresses the importance of disinfect-ing hard surfaces, regular hand-washing 'with soap and warm water' and avoiding crowded gatherings, 'especially in enclosed spaces.' And just as in 1918 doctors debated the pros and cons of face masks, so in 2008 the department of health argues that 'wearing masks at all times is not practicable' and, in any case, there is 'little actual evidence of proportion-ate benefit from widespread use.' It concludes that while face masks may be appropriate for doctors and nurses who are exposed to infected patients, the government will 'not be stockpiling face masks for general use,' though there is nothing to stop individuals from buying their own should they so desire.

Depressingly, the one area where Davies foresees a need for a significantly different approach to 1918 is on the level of community action. In 1918, women spontaneously came to each other's aid calling on neighbours to check if they needed food and medicines and looking after their children if they were unable to care for them. Today, Davies is hoping to foster a similar community spirit by asking people to 'Be a flu friend' and identify someone in their neighbourhood now who they would look out for in the event of a pandemic. She has also recruited the Women's Institute and the tissue manufacturer Kleenex to help spread the word that people should use tissues to stifle coughs and sneezes wherever possible – an initiative known as the 'tissue mission.' The big fear of course is that as with the run on the Northern Rock bank in 2007, in a crisis all sense of community will evaporate and it will be every Briton for him or herself. Does Davies fear for Britain in the event of a pandemic or does she foresee a miraculous return of the 'Blitz spirit'?

'Both,' she says after a pause. 'I hope that people will remain sensible and calm and only ask for antivirals when they need them but I think that the media are also going to be important because they can encourage people to do the decent thing.'

***

Hypothetical scenarios are all very well and good but what would happen in the event of a real alert? One person who has a good idea is Petra Graf. An award-winning Swiss camerawoman based in London, Graf is no stranger to life-threatening situations. Travelling the world's trouble spots shooting documentaries for the BBC and Channel 4, Graf has braved mortar attacks in Iraq, filmed would-be Cuban assassins in Florida, and hung out with teenage prostitutes on the streets of London. But Graf's scariest assignment to date came in June 2006 when WNET in New York asked her to fly to Vietnam to make a documentary about H5N1. Like most people at the time Graf knew little about the virus other than what she'd read in the newspapers. Fortunately, her New York producer was better prepared and had stocked up on Tamiflu and Relenza. He also brought face-masks, latex gloves, rubber boots and goggles for each member of the team to wear whenever they were filming on infectious disease wards or in slaughterhouses. The producer was very strict. 'He warned us not to take off our masks for any reason as new human cases were being regularly reported around the country,' Graf recalls.[230]

During the three-week shoot, Graf travelled from the Red River delta in the north to the Mekong Delta in the south filming at chicken farms and poultry markets. Then, on the final day she visited the Bach Mai hospital in Hanoi where Sy Tuan, the young Vietnamese man who had lain mortally ill on a ventilator the previous year, had returned for a check-up with Dr Van. Sy Tuan had made a full recovery but during the filming of the interview Dr Van informed Graf that she had four other cases of H5N1 and seven suspected cases on her ward. However, what worried Graf even more was her visit to a nearby slaughterhouse.

'We saw them cut the throat of a duck. The birds were on a carousel hanging upside down and were still fluttering and moving around while I was filming. I got splattered by some of the birds' blood.'

On the last day of shooting – 24 June – Graf began to feel unwell, but thinking it was just a cold brought on by the 40° heat and all the moving around between air-conditioned offices she decided she was fit enough to fly home. On the return flight to London, however, Graf started running a temperature and coughing and by the time she reached her flat in Hackney she was feeling seriously unwell. That afternoon she called the New York production office for advice and was told that in view of where she'd just been she should consult her doctor. Alarmingly, Graf's GP had never heard of Tamiflu. Nevertheless, she had the sense to call the local hospital in Homerton for advice. According to Graf the hospital had no idea what to do either and told her doctor 'don't come anywhere near us.' Instead, Graf was sent back to her flat in Hackney and told to await further instructions.

The next few hours would provide the first and to date only live test of the Department of Health's pandemic response system.

'Within half an hour of arriving home I had one phone call after another,' says Graf. 'I spoke to two or three different doctors including people from the health department. They told me I'm not allowed to speak to anybody and to wait at home, someone would come up and pick me up. I waited and waited for hours.'

Eventually, at around 2am Graf's doorbell rang and she opened the door. Standing on her doorstep were two medics wearing full biohazard suits. Ordering Graf to don a third biohazard suit, they apologized for the delay – apparently it had taken them some time to locate the key to the cupboard where the suits were stored. In all, Graf had been kept waiting for seven hours.

Graf was rushed by ambulance direct to the Royal Free Hospital in Hampstead where she was wheeled into a special containment unit via a side entrance. 'It was quite terrifying. They told everyone to keep away from me.'

There, she spent the next three days in isolation sleeping under disposable paper sheets, wearing disposable paper gowns and being tended to by a steady stream of specialists. 'I wasn't worried the first day but after the second day, with everyone so careful and anxious, it began to occur to me that I might really be infected with the virus.'

By now the Health Secretary Patricia Hewitt was receiving regular updates and the media were camped outside the hospital. Anyone wishing to visit Graf had to pass through a decontamination room. However, though the biosecurity around Graf was tight no one had thought to extend the measures to her flatmate who worked at the BBC and who had already visited the corporation's offices at White City several times.

In the event, Graf's illness proved to be a false alarm. She had an ordinary case of flu, not H5N1, and on the fourth day, with her temperature returning to normal, she was discharged. In the end Graf hadn't even needed oseltamivir, just aspirin for her fever.

Today, she says, the experience seems like a surreal nightmare. 'It was bad timing. I've done much more dangerous things. I've been in Baghdad and God knows where. It was just bad luck.'

Was she reassured by the authorities' response? 'To be honest I was a bit surprised by how long it took for the medical team to get to my house, and all because they couldn't find a key to the cupboard where they keep the suits. Hopefully, when there's a real pandemic they'll be better prepared.'

It is hard not to share Graf's concerns. What if she had had H5N1 and had infected her flatmate, and what if he had gone on to infect the BBC? Within days the whole of the corporation would have been down with flu and, given the BBC's key role as a public broadcaster, the incident could have triggered nationwide panic. On the other hand, Graf was eventually taken to an isolation ward where by her own account she received excellent care. Moreover, in the event of a real pandemic the department of health would almost certainly have several days forewarning, giving it plenty of time to activate its plans and ensure that local health service providers were in a position to coordinate their responses. Having said that, in the event of a rerun of 1918 no one can predict how the NHS will cope which is why Davies hopes that the government's pandemic preparedness plans will never be activated in earnest.

'We need to be honest about how difficult things are going to be,' she acknowledges. 'Besides if I said no, everything's going to be fine, the health service will manage, you wouldn't believe me.'

epilogue

'The disease simply had its way. It came like a thief in the night and stole treasure.'

– George Newman, *Ministry of Health Report on the Pandemic of Influenza*

In March 1919, at the height of the third wave of the pandemic, *The Times* carried a brief item enumerating the sums paid out by the Prudential Assurance Company in connection with the pandemic. In all, since the previous November, claims for deaths from influenza had cost the Pru more than £650,000. This compared to just £279,000 paid out to widows of men killed in action. However, according to *The Times*, what shocked the Pru's board was not so much the amount as the fact that the mortality had occurred mainly between the ages of 16 and 40.[231]

For all the advances in virology and our superior understanding today of the pathology and immunology of influenza, this W-like peak observed in young adults has never been satisfactorily explained. Nor, nearly a century later, are scientists in a much better position to say whether the Great Flu was unique – part of the never-to-be-repeated ecology of 1918 – or whether it was just bad luck that the virus hit a world reeling from nearly five years of warfare.

Reviewing the pathology and pattern of deaths in his introduction to the Ministry of Health's voluminous 1920 *Report on the Pandemic of Influenza*, George Newman considered several

Figure E.1 George Newman
Credit: Wellcome Library, London

explanations for the high mortality observed in young adults. First, he wondered if the phenomenon might be due to older members of the population enjoying immunity to the disease because they had survived the Russian flu. In fact, we now know that the virus that triggered the Russian flu was almost certainly an H3. Nevertheless, if the serological and epidemiological evidence that points to the previous 1847 strain being an H1 is accurate then Newman's hypothesis may well merit further investigation. On the other hand, even writing in 1920 Newman had doubts that such a theory could provide a complete explanation, pointing out that any immunity people may have acquired from having survived a previous pandemic appeared, from the obtainable evidence, to be 'slight, transient, variable and incomplete.'[232]

Next, Newman considered the theory that the change in age incidence was due to the disruption of populations as a result of war.

Soldiers have been aggregated for war purposes, young men and women in munition works, large sections of adult populations have moved in bulk owing to trade or transport exigencies, and thus the disease had greater opportunity of fastening upon these aggregated populations under exceptional surroundings.[233]

This explanation has immediate intuitive appeal. The war provoked unprecedented mass movements of people – Chinese labourers and Commonwealth troops to northern France, Americans from the Midwest to Brest and Bordeaux, women quitting dull domestic positions in Surrey and Kent to join their sisters in munitions factories in London and the Midlands. Then there was the constant passage of soldiers to and from the Front in confined train carriages – conditions that provided the perfect conditions for the transmission of the virus. However, while it is possible that the crowding together of large numbers of young men in trains, army camps and on naval vessels may have enhanced the virulence of the virus, the problem with this explanation is that the same patterns of influenza morbidity and morality were seen in *all* countries and nations in 1918, including those not party to the war where young adults were not subject to the same social disruption.

A third explanation considered by Newman was that the adolescent and adult populations were disproportionately affected by the stresses and strains of war. Once again, this has immediate intuitive appeal. The pressure on young men to enlist and do their 'duty' was intense, to the extent that men seen out of uniform risked being handed white feathers and being branded 'shirkers.' For these raw recruits it was their first time away from home and their first experience of trench warfare. Moreover, for those left behind, whether because they were physically unfit or their occupation was deemed essential to the war effort, the pressure to 'carry on' and make a contribution was intense. But while it is probable that many young people continued working in spite of illness – or returned to work before they were fully recovered from the flu – once again this explanation fails for the same reason as the previous one, namely that the virus also decimated remote populations in parts of the world that were not subject to the same stresses.

The fourth and final explanation considered by Newman was that the 1918 virus was one to which 'adolescent and adult tissues were particularly susceptible.' In particular, he noted the heliotrope cyanosis and extensive haemorrhaging of the lungs. However, while this pathology was characteristic of the severe or 'pneumonic' cases, in the vast majority of cases there were no unusual complications. Moreover, while today virologists have a better understanding of viral invasions of the respiratory tract, the theory that the haemorrhaging and cyanosis was due to an extreme immune response – the so-called cytokine storm – remains just that, a theory.

Next, Newman considered what role, if any, social conditions might have played. Were the poor or those living in overcrowded conditions in the great towns more likely to contract the disease than the rich and well-off? Once again, the evidence was inconclusive. In reviewing the block censuses that had been carried out in Newcastle, Major Greenwood found a small correlation between overcrowding in tenements and attack rates, but the block censuses in Manchester and Leicester produced no such correlations. Furthermore, when the Registrar General compared mortality across 29 London boroughs with health and wealth indicators, such as the number of domestic servants living in a household, it found only a moderate association.[234] Influenza, it concluded, had fallen equally on 'the sanitarily just and unjust.' Subsequent studies from other countries and cities have proved similarly contradictory and inconclusive – in Paris, for instance, the flu hit the affluent seventeenth *arrondissement* harder than the poorer and densely populated thirteenth.[235]

Newman was not alone in being puzzled by the democratic nature of the flu or the unusual mortality in young adults. William Hamer, the Chief Medical Officer of the London County Council, had also been intrigued by the W-shaped mortality curve. However, in addition to the mortality peak observed in young adults Hamer noted a high incidence of deaths in children aged 2–5. The explanation, he suggested, was that the disease circulating in the Army camps was of a more 'severe type' and that when men in service had returned home they had transmitted the infection to their families. Another factor was the restricted wartime diet which, he argued, impacted growing children and young adults working long hours harder than older, more sedentary members of the population. If

Hamer is correct, this might explain why Ada Darwin's parents and her baby brother Noel died but she survived.

However, it was when they turned their minds to the origins of the pandemic and the epidemiology of the outbreaks that Newman and Hamer came closest to agreement. Noting the 'smouldering' of influenza in England and Wales in 1915 and the similarity between the cases of heliotrope cyanosis and 'purulent bronchitis' observed at Aldershot and Etaples in 1916–17, Newman argued that it was 'impossible to escape the conclusion that these various conditions bore a fundamental relationship to each other and to the pandemic of influenza.'

Hamer was even more convinced of a connection and tried to reconstruct the disease outbreaks with a view to discerning a pattern. He dismissed the notion of a Spanish origin but was not convinced that the pandemic had started at Etaples either. Instead, he believed the pandemic had begun on the eastern seaboard of the US with the first reports of a 'mild grippelike disease' in March and April of 1918.[236]

In the end, however, neither Hamer nor Newman were able to reach a definite conclusion. Instead, they took refuge in metaphor. For Hamer the pandemic could be compared to a pond into which someone had thrown a stone, so that 'from the big centres of population the waves of influenza travel outwards to remote parts of the earth and then, later, are in due course reflected back.' In Newman's case the origins lay in a Darwinian struggle or a 'contest' as he put it between man and 'germ.' Ironically, it was in the congestion of public transport and the growth of public assemblies and entertainments – hallmarks of advanced Western civilizations – that 'a strategical advantage was given to the enemy.' He concluded:

> Finally, in the provision of countless incubators, whether in garrisons, war-time factories, or abnormally over-crowded and ill-ventilated

means of transport and places of entertainment, the opportunity was afforded for the development of destructive powers which secured to the enemy a decisive and overwhelming victory.[237]

Today you could argue that from the wet markets of Guandong to the slums of Dacca and the poultry farms of Arkansas, those incubators are even more numerous. Only time will tell whether they will cede the virus a similarly decisive and overwhelming victory.

biographical postscript

George Newman

A lifelong Quaker with a strong philosophy of public service, Newman had a unique vision of how the state should expand its role in medicine. Appointed Chief Medical officer of the Board of Education in 1907, Newman became convinced that preventive medicine should be central to the government's agenda, arguing that the best way to improve the nation's health was to educate people to help themselves. Aware that he could only advance his agenda from a position of authority, he was an early supporter of the new Ministry of Health, forming a close personal alliance with Christopher Addison, its Minister elect, and Robert Morant, the key architect of Lloyd George's national insurance reforms.

In 1919 Morant helped engineer Sir Arthur Newsholme's retirement from the Local Government Board, thus paving the way for Newman to succeed him as CMO. Morant liked to joke that he and Newman had a 'Siamese twin arrangement' – a reference to Morant's pre-civil service days when he had served as tutor to the crown prince of Siam. The men had been looking forward to renewing this arrangement at the Ministry of Health, where Morant was ensconced as Permanent Secretary, but it was not to be: in March 1920, as influenza once again elevated the death rate in British towns and cities, Morant contracted pneumonia dying very suddenly at his home in Thurloe Square, Knightsbridge.

Although flexible and pragmatic in medical matters, Newman was uncomfortable in Whitehall and had looked to Morant for advice in civil service matters. Morant's unexpected death robbed him of that guiding hand and when in 1921 Addison was also forced to resign from government Newman found himself doubly adrift. The result was that although Newman was to serve as CMO for a further 14 years and proved an effective advocate of public health, particularly in the area of personal hygiene, he failed to affect any further major reforms in medical administration. As the *BMJ* concluded in Newman's obituary in 1948, he 'was always the medical man rather than the Civil Servant, and the health propagandist above all.'[238]

Walter Morley Fletcher

Morant's death also came as a blow to Fletcher. It was Morant who, following Fletcher's appointment to the MRC, had fired his passion for scientific research in the public interest, and after the war Morant had lobbied for the MRC's independence from the Ministry of Health, ensuring that it answered directly to a ministerial committee of the Privy Council. Thanks to this arrangement and a post-war budget of £125,000 a year, Fletcher was able to push on with research unencumbered by political considerations and in the 1920s he acquired a site at Mill Hill, north London, where he was able to build the Farm Laboratory – what would later become the National Institute for Medical Research – dedicated to the elucidation of diseases such as diabetes, rickets and influenza.

Although Fletcher had escaped what his wife, Maise, writing in the 1950s called the 'appalling ravages of the black influenza,' he remained 'terribly concerned' about it and 'from then on initiated a real attack on the disease.'[239] In 1933, the attack paid off when a team of researchers at the Farm Laboratory headed by Sir Patrick Laidlaw took throat washings from colleagues infected with influenza, filtered them to remove any bacteria and succeeded in infecting ferrets, thus proving that influenza was caused by a filter-passing virus. Soon after, one of the ferrets sneezed in the face of a scientist who was handling it and the scientist went on to develop flu, thus completing the proof.

In 1935 one of the same researchers discovered that the flu virus could be cultivated in chick embryos in very large numbers, paving the way for today's mass production of flu vaccines.

Major Greenwood

In 1919 Major Greenwood joined Newman at the Ministry of Health where he took charge of the division of medical statistics. Nine years later, in 1928, he was elected professor of epidemiology at the London School of Hygiene and Tropical Medicine, a position he held until his retirement in 1945.

Although at the 'Discussion on Influenza' at the Royal Society of Medicine, Greenwood had agreed with Newsholme that the second wave could not have been predicted 'with certainty,' following the war he and Newsholme increasingly found themselves at loggerheads. On the surface their differences revolved around the issue of tuberculosis and how best to eradicate it, with Newsholme favouring 'stamping out' measures directed against the bacillus and Greenwood preferring to emphasize predisposing environmental factors. But at the root of their antipathy was Greenwood's advocacy of the latest biometrical methods and his intolerance of those who, like Newsholme, he believed lacked the mathematical training to appreciate them.

Interestingly, for all that Greenwood was keen to see epidemiology be taken seriously as a science, he remained wedded to older holistic notions of disease arguing that the problem of disease as a mass phenomenon could not be reduced to the 'chasing of the "germ" or the discovery of a "vaccine."' A keen student of history, Greenwood was said to have amazing powers of recall and could quote both Virgil and Marx at will. However, he was also prone to fits of depression and towards the end of his career he surprised many of his colleagues by becoming a staunch defender of psychoanalysis.

William Hamer

An idiosyncratic epidemiologist and physician, Hamer joined the London County Council in 1892 after training at St Bartholomew's hospital. Three years later, aged just 33, he became a fellow of the Royal College of Physicians and in 1912 he succeeded Sir Shirley Murphy as the Chief Medical Officer of the LCC.

Elected President of the Epidemiological Section of the Royal Society of Medicine in 1915, Hamer, like Greenwood, did not share the confidence of bacteriologists that the germ theory of disease provided a complete explanation of the aetiology and epidemiology of epidemics. In particular, he pointed out that bacteriology was unable to explain why influenza epidemics came and went, why some people got sick and others did not, and why the disease appeared to 'cling' to particular places or struck only in certain seasons of the year. Instead, Hamer looked to the older and hazier notion of an 'epidemic constitution' – an environmental conception of disease that harked back to the ideas of the 18th century British physician Thomas Sydenham.

It was Hamer's interest in epidemiology that led him to focus on a cluster of fatal pneumonias in the winter of 1917–18 at a boys' reformatory school in London – outbreaks that in retrospect he would come to regard as 'trailers' of the pandemic. Hamer observed that although the cases had been spread over several months the boys' illnesses had proved very rapidly fatal and had occurred just before the 'first distinct rise' in London's overall death rate. Writing in 1919, he concluded that 'something new and special in the behaviour of the influenza... had been definitely aroused.'[240]

Sir Arthur Newsholme

After resigning from the LGB in 1919 to avoid serving under Newman at the Ministry of Health, Newsholme went on to enjoy a very active retirement lecturing on public health administration at the newly created School of Hygiene and Public Health at John Hopkins University and conducting a major study of European healthcare systems for the Milbank Fund. In 1932 he and the secretary of the fund visited the Soviet Union to study socialized health care, later co-authoring a book entitled 'Red Medicine.' Newsholme was a prolific writer and he spent the last ten years of his life writing a two-volume history of public health. He died at home in 1943 at the age of 86.

Although Newsholme had been an exemplary medical health officer, using his time at Brighton to stamp out a range of preventive diseases, including tuberculosis and scarlet fever, he had a sanitarian mindset that made him suspicious of bacteriology and the contribution that contagionist theories of disease could make to public health. Coupled with his passion for local municipal control, this arguably blinded him to the wider opportunities offered by state intervention in medicine, not least of all in the LGB's lackadaisical response to the influenza epidemic.

James Niven

James Niven served as Medical Officer of Health for Manchester until 1922. When he took up the post in 1893 he had just two clerks. By the time he retired he had charge of 860 staff, two sanatoriums, an isolation hospital and 12 maternity and child welfare centres. Not only that but thanks to his innovations (he was one of the first MOHs to propose the notification of tuberculosis and pioneer city-wide health visitor services) Manchester's reputation for salubrity soared.

According to Newman, Niven had a reputation as a deep and original thinker and his advice was constantly sought. 'Again and again those who believed they had broken fresh ground found the problems stated and their solution proclaimed in Niven's writings of years before.'[241]

But though during his time in office Niven had trained many of the country's public health officers, he found it difficult to make friends and following his retirement he grew increasingly despondent (his wife had died in 1912 and his daughters had long since left home). At first Niven tried living alone in lodgings, but on 28 September 1925 he travelled to the Isle of Man and checked into a hotel. Two days later his body washed up at Onchan harbour.

At the subsequent inquest, the coroner concluded that Niven had committed suicide by taking poison and drowning after swimming out to sea. He was cremated at Manchester on 7 October 1925.[242]

notes

1 N.P. Johnson and J. Mueller, 'Updating the accounts: global mortality of the 1918–1920 "Spanish" influenza,' *Bulletin of the History of Medicine*, 76: 1 (Spring 2002), 105–15.

2 *The Times*, 18 December 1918.

3 *Independent*, 22 October 2005.

4 N. Johnson, *Britain and the 1918–19 influenza pandemic: a dark epilogue* (London: Routledge, 2006).

5 R. Collier, *The Plague of the Spanish Lady: the influenza pandemic of 1918–1919* (London: Macmillan, 1974).

6 R. Porter, 'The Patient's View: Doing Medical History From Below', *Theory and Society* (March 1985) 14, 2: 175–98.

7 Johnson (2006), 180.

8 M. Greenwood, *Epidemics and crowd-diseases: an introduction to the study of epidemiology* (London: Williams & Norgate, 1935), 326.

9 F.G. Crookshank, *Influenza: essays* (London: William Heinemann Medical, 1922).

10 Letter to Susan Owen, June 24 1918, in H. Owen and J. Bell (eds), Wilfred Owen Collected Letters (London: Oxford University Press, 1967).

11 Ibid.

12 'Overarching Government Strategy to Respond to an Influenza Pandemic: Analysis of the Scientific Evidence Base', Civil Contingencies Secretariat, Cabinet Office (November 2007), 22–3.

13 Today, given the growth of the UK's population, it is estimated that a virus with a similar attack rate would kill 375,000. If the clinical attack rate was a higher 35 per cent with the same 2.5 per cent fatality rate the dead could number more than half a million. See 'Pandemic flu: a national framework for responding to an influenza pandemic', *Cabinet Office and Department of Health* (22 November 2007), 27.

14 C. Creighton, *A history of epidemics in Britain* (Cambridge: The University Press, 1894) 1: 237–71.

15 J. Caius, *A boke, or counseill against the disease commonly called the sweate, or sweatyng sicknesse* (London: Richard Grafton, 1552).

16 W. Beveridge, *Influenza: The Last Great Plague, an unfinished story of discovery* (London: Heinemann, 1977), 25.

17 Creighton (1894) 2: 338.

18 For instance, during a particularly bad influenza outbreak in 2000 the distinguished astronomer Fred Hoyle revived the 'germs from space' theory by suggesting that influenza viruses might hitch a ride to earth in the carbonaceous particles of comets.

19 Beveridge (1977), 24–5.

20 Crieghton (1894) 2: 348.

21 D.T. Thomson and R. Thomson, *Annals of the Pickett-Thomson Research Laboratory* (London: Bailliere Tindall & Cox, 1934), 1: 47.

22 Ibid, 1: 7.

23 A.C. Lowen, *et al.* 'Influenza Virus Transmission Is Dependent on Relative Humidity and Temperature', *PloS Pathogens*, 3: 10 (2007): 151.

24 *H. influenzae* was the first living organism to have its entire genome sequenced. Commonly found in the nasal passage and respiratory tract it is usually harmless. However, in infants *H. influnezae* type b (Hib) can trigger *bacteremia* and acute bacterial *meningitis*, and in association with viral infections or when the immune system is otherwise weakened it can also trigger respiratory tract infections such as pneumonia.

25 'The Progress of the World', Review of Reviews, 1: 2 (February 1890), 87.

26 S. West, 'The Influenza Epidemic of 1890 as experienced at St Bartholemew's Hospital and the Royal Free Hospital', *St Bartholemew's Hospital Reports*, 26 (1890).

27 H. Littlejohn, *Report of the Medical Officer of Health for Sheffield*, 2 November, 1891.

28 53rd, 54th, 55th and 56th Annual Report of the Registrar-General of Births, Deaths and Marriages in England (London: HMSO, 1890–94).

29 H.F. Parsons, 'Further Report and Papers on Epidemic Influenza 1889–92', *Local Government Board* (London: Eyre & Spottiswoode, 1893).

30 A.W. Crosby, *America's forgotten pandemic: the influenza of 1918* (Cambridge: Cambridge University Press, 2003).

31 *The Times*, 25 June 2006.

32 Interestingly, when men first reported ill with trench fever it was termed 'pyrexia of uncertain origin' or P.U.O., the same generic label that would be applied to the initial cases of Spanish influenza.

33 T.J. Mitchell and G. M. Smith, *Medical Services: casualties and medical statistics of the Great War* (Imperial War Museum, 1977).

34 V. Brittain, *Testament of youth: an autobiographical study of the years 1900–1925* (London: Virago, 1978).

35 D. Gill and J. Putkowski, *Le Camp Britannique d'Etaples. The British base camp at Etaples, 1914–1918* (Musee Quentovic, Ville d'Etaples, 1997).

36 Ibid.

37 Major General Charles Foulkes, quoted in Nick Howard, 'Chemical Weapons, Mass Mobilisation & the First World War Influenza Pandemic

of 1918–19'. Paper delivered to 10th Annual conference on Alternative Futures and Popular Movements at Manchester Metropolitan University, 30 March–1 April 2005. I am indebted to Nick Howard for his insights into the use of mustard and other gases on the Western front. Howard points out that British munitions workers who handled phosgene gas missiles also suffered higher than normal incidences of broncho-pneumonias and flu. The American military also conducted explosive tests with poison gas shells in Washington and Baltimore within sight of trolley car commuters.

38   Private Clifford Saunders, Imperial War Museum.

39   J. Oxford, *et al.*, 'A Hypothesis: the conjunction of soldiers, gas, pigs, ducks, geese and horses in Northern France during the Great War provided the conditions for the emergence of the "Spanish" influenza pandemic of 1918–19', *Vaccine* 23 (2005), 940–5.

40   Interview with author, March 2008.

41   J.A.B. Hammond, William Rolland, T.H.G. Shore, 'Purulent Bronchitis: a study of cases occurring amongst the British troops at a base in France', *Lancet* (14 July 1917) 190: 41–6.

42   Letters from Sir John Rose Bradford and his wife Mary Bradford, largely while serving with the Royal Army Medical Corps in France, 1905–1921. PP/BAR/Z/9, Contemporary Medical Archive Centre, Wellcome Library.

43   *Proceedings of the Royal Society of Medicine*, 1918–19, 12: 97–102

44   A. Abrahams, *et al.*, 'Purulent Bronchitis: its influenzal and pneumococcal bacteriology', *Lancet* (8 Sept 1917) 190: 377–82.

45   Harold Owen and John Bell (eds), *Wilfred Owen, Collected Letters* (London: Oxford University Press, 1967), p. 421. Letter to Susan Owen, 1 January 1917.

46   Ibid, p. 521. Letter to Susan Owen, 31 December 1917.

47   Brittain (1978).

48   'The Influenza,' Winston Churchill, Harrovian school newsletter, 10 December 1940.

49   C. Playne, diary entry 12 February 1918. *Caroline Playne Collection* – MS1112, Senate House Library, University of London.

50   Kipling's grief over Jack's death has been immortalized both in poetry and in a recent book and ITV television drama.

51   A.G. Burn diary, Imperial War Museum. Quoted in T. Wilson, *The Myriad Faces of War* (Cambridge: Polity Press, 1996), 507.

52   *The Times*, 17 December 1917.

53   In the play, Eliza Doolittle remarks: 'My aunt died of influenza, so they said. But it's my belief they done the old woman in...Why should she die of influenza when she come through diphtheria right enough the year before?'

54   J. Winter and J.L. Robert, *Capital cities at war, London, Paris, Berlin 1914–19* (Cambridge: Cambridge University Press, 1997), 318–20. J.M. Winter, *The Great War and the British people* (London: Macmillan Education, 1986).

55   *The Times*, 13 December 1917.

56   J.M. Barry, *The Great Influenza: the Epic Story of the Deadliest Plague in History* (New York: Viking, 2004), 95.

57   C.R. Byerly, *Fever of war: the influenza epidemic in the U.S. Army during World War I* (New York; London: New York University Press, 2005), 70. L. Iezzoni and D.G. McCullough, *Influenza 1918: the worst epidemic in American history* (New York: TV Books, 1999), 21–4.

58   Barry (2004), 91–5. G. Rice and L. Bryder, *Black November: the 1918 influenza pandemic in New Zealand* (Christchurch, New Zealand: Canterbury University Press, 2005), 50–1.

59   Brittain (1978).

60   The appellation was not meant to impute an Italian origin but was chosen because the song about the soldier of Naples was said to be as 'catchy' as the disease. See B. Echeverri, 'Spanish Influenza seen from Spain', in H. Phillips and D. Killingray, *The Spanish influenza pandemic of 1918–19: new perspectives* (London; New York: Routledge, 2003).

61   Indeed, there is evidence that influenza may have been similarly widespread in Britain around the same time. In June the *BMJ* carried a report saying that in May there had been 'a severe outbreak of influenza in a certain district of this country' without specifying where and drawing a parallel between the sudden appearance of the disease and the outbreak in 1889–90 which heralded the first wave of Russian flu. However, in almost the same breath the journal dismissed comparisons to the 'plague of 1889' saying that the low mortality rate rendered 'alarmist suggestions premature' and the probability of an influenza pandemic should be discounted 'before a scare of cholera is raised'. See *BMJ* (1 June 1918), 1: 627.

62   Reports on Public Health and Medical Subjects, No. 4, Ministry of Health 'Report on the Pandemic of Influenza 1918–19' (London: HMSO, 1920).

63   Barry (2004), 171.

64   Caroline Playne diary, 24 March and 31 May 1918.

65   *The Times*, 25 June 1918.

66   O.H. Gotch and H.E. Whittingham, 'A report on the "Influenza" Epidemic of 1918', *BMJ* (27 July 1918), 2: 82–5.

67   Report on the Pandemic (1920).

68   W.J. Wilson and P. Steer, 'Bacteriological and Pathological Observations of Influenza in France during 1918', National Archives: FD 5/187. The theory is not as far-fetched as it seems: according to Professor Peter Openshaw, an expert in respiratory medicine at Imperial College,

London, in the same way that the influenza virus can prepare the way for secondary bacterial infections of the lungs so poison gas, by provoking a massive pulmonary assault, can also encourage bacterial infections.

69 Letter, W. Fletcher to R. Reece (6 November 1918), 'Influenza Committee – Correspondence with Local Government Board and War Office 1918–20'. National Archives: FD 1/535.

70 For further discussion of Lloyd George's strategy for reforming health care and the negotiations which led to the establishment of a Ministry of Health see F. Honigsbaum, 'The Struggle for the Ministry of Health', Occasional Papers on Social Administration, No. 37 (Social Administration Research Trust, 1970).

71 'Discussion on Influenza,' *Proceedings of Royal Society of Medicine* (1919), 12: 1 and 2.

72 *The Times*, 25 June 1918.

73 Discussion on Influenza (1919).

74 *Glasgow Herald*, 17 and 24 July 1918.

75 *Yorkshire Telegraph*, 3–5 July 1918.

76 *Glasgow Herald*, 23 July 1918.

77 *Salford Reporter*, 29 June 1918.

78 *The Times*, 25 June 1918.

79 For further discussion see chapter six and epilogue.

80 'London Letter', 20 August 1918, *JAMA* (July–September 1918), 71: 990.

81 *Manchester Guardian*, 12 September and 14 September 1918.

82 *Manchester Guardian*, 14 September 1918.

83 *Manchester Guardian*, 13 September 1918.

84 Crosby (2003), 152.

85 *Manchester Guardian*, 23 September 1918.

86 John Grigg, *Lloyd George: War Leader* (London: Allen Lane, 2002).

87 Crosby (2003), 37–8.

88 R.N. Grist, 'Pandemic Influenza 1918', *BMJ* (22–29 December 1979), 1632–3.

89 It is not known if Roy survived the flu. His letter was found in a trunk of old medical papers in Detroit in 1959 and handed to the department of epidemiology at the University of Michigan from where it was forwarded to a relation, a professor of infectious disease in Glasgow.

90 Byerly (2005), 79.

91 Ibid, 102.

92 Ibid, 103.

93 Ibid, 97–8.

94 *Manchester Evening News*, 7 October 1918.

95 *Manchester Evening News*, 2 November 1918.

96 *Glasgow Herald*, 23 October 1918.

97   *Yorkshire Telegraph,* 19 October 1918.

98   *Daily Mirror,* 22 and 23 October 1918.

99   Report on Pandemic (1920).

100   *The Times,* 28 October 1918.

101   *Yorkshire Telegraph,* 19 October 1918.

102   *Manchester Evening News,* 23 October 1918.

103   Wilson (1986), 650.

104   M. Phillips and J. Potter, *A Sheffield Munitions Worker: Septimus Bennett, Artist in Arms, 1915–18* (Durham: Portland Press, 2001).

105   *The Times,* 2 February 1921.

106   Crosby (2003), 322.

107   Ibid, 319.

108   Bertram E. Copping, Letter 18 June 1973. Collier Collection.

109   A.M. Forbes, Letter 11 May 1973. Collier Collection.

110   F.W.P. Frewer, Letter 11 May 1973. Collier Collection.

111   Dorothy E. Jack, Letter 5 May 1973. Collier Collection.

112   Edith Dilks, Letter 9 May 1973. Collier Collection.

113   Maurice E.M. Jago, Letter 27 June 1973. Collier Collection.

114   *Daily Mirror,* 24 October 1918.

115   *Daily News,* 10 October 1918.

116   'Memorandum on Epidemic Catarrhs and Influenza', Local Government Board, October 1918. *The Times,* 22 October 1918.

117   *Manchester Evening News,* 23 October 1918.

118   *The Times,* 23 October 1918.

119   In 1918 a virus was defined as submicroscopic infectious organism that could be filtered but not grown in vitro. In the 1880s Pasteur had shown that rabies was a 'filter-passer' and by the outbreak of WWI a number of other diseases had also been shown to be due to viruses. However, although scientists knew that repeated passage of serums containing these filter-passing organisms could induce immunity against disease, they continued to conceive of viruses as essentially ultramicroscopic bacteria that multiplied by binary fission rather than, as we now know, by invading host cells and replication.

120   Letter, Fletcher to Newsholme (23 October 1918), 'Influenza Committee – Correspondence with Local Government Board and War Office, 1918–20'. National Archives.

121   Letter, Fletcher to Waldorf Astor (29 October 1918), Addison Papers, Oxford, Bodleian Library, Box 21.

122   Diary of Sir George Newman. MH 139/3, National Archives.

123   *Manchester Guardian,* 30 October 1918.

124   Letter, Eva Keating, 5 May 1973. Collier Collection.

125   *The Times,* 31 October 1918.

126   *Manchester Evening News,* 23–24 October 1918.

127 J. Niven, 'Report on the Epidemic of Influenza in Manchester 1918–19', in Report on Pandemic (1920), 479–80.
128 Ibid, 480.
129 *Manchester Guardian*, November 12 1918.
130 Ibid.
131 *Manchester Guardian*, November 13 1918.
132 *Manchester Evening News*, November 12 1918.
133 Report on Pandemic (1920), 474–7.
134 Interview with author, 5 November 2005. Letter, 2 September 2007.
135 W. Hamer, 'Report on Influenza' (June 1919), LCC official publication, 210: 1963, 2.
136 Report on Pandemic (1920), 499.
137 'Our present Knowledge of Epidemic Catarrh', *Lancet*, 26 October 1918, 559.
138 'Bacteriology of "The Spanish Influenza",' *Lancet* (10 August 1918), 177. 'Prevention and Treatment of Influenza', *BMJ* (16 November 1918), 546.
139 'Royal Society of Medicine Discussion of Influenza', *BMJ* (30 November 1918), 574–5.
140 Letter, W. Camac Wilkinson, *BMJ* (16 November 1918), 551.
141 J. Niven, *Observations on the history of public health effort in Manchester* (Manchester: J. Heywood, 1923).
142 'The utilisation of vaccine for the prevention and treatment of influenza', *Lancet*, 26 October 1918, 565.
143 Johnson (2006), 144.
144 Letter, Fletcher to Cummins (6 December 1918), Influenza Research. National Archives: FD 1/529.
145 Letter, Fletcher to Gibson (2 December 1918), Influenza Research. National Archives: FD 1/529.
146 Letter, Fletcher to Newsholme (23 October 1918), Influenza Committee Correspondence with Local Government Board and War Office 1918–20. National Archives: FD 1/535.
147 Medical Research Committee, 'Studies of Influenza in hospitals of the British Armies in France 1918.' *Special Report Series* (London: HMSO, 1919), No. 36.
148 Discussion on Influenza (1919).
149 Report on Pandemic (1920), 353. K. McCracken and P. Curson, 'Flu downunder: a demographic and geographical analysis of the 1919 epidemic in Sydney, Australia', in Phillips and Killingray (2003), 110–31.
150 Johnson (2006), 123.
151 PBS online 'The American Experience: Influenza 1918'.
152 Collier (1974), 144–5.

153 *Sun-Telegram*, 12 Jan 1970.
154 'Exploding the Myths, White Paper Pandemic Influenza', *Aon* (December 2006), 15.
155 Johnson (2006), 144.
156 *Daily Sketch*, 4 January 1919.
157 *The Times*, 16 January 1919.
158 *The Times*, 31 January and 18 February 1919. W. Hamer, 'Report on Influenza', LCC official publication (June 1919) 210: 1963, 2.
159 R. Graves, *Goodbye to all that: an autobiography* (London: Jonathan Cape, 1929).
160 Johnson (2006), 45–6.
161 B. Hood, 'Notes on St. Marylebone Infirmary (later St. Charles Hospital) 1910–1941'. Contemporary Medical Archives Centre, Wellcome Library: GC/21.
162 B.E. Copping, Letter 18 June 1973. Collier Collection.
163 P. Zylberman, 'The Great War and the 1918 Spanish Influenza epidemic in France', in Phillips and Killingray (2003).
164 Report on Pandemic (1920).
165 J.M. Winter, *Sites of Memory, Sites of Mourning: the Great War in European cultural history* (Cambridge; New York: Cambridge University Press, 1995), 18–22.
166 Crosby (2003), 194–6.
167 C.S. Sykes, *The Big House: The Story of a Country House and its Family* (London: Harper Collins, 2004).
168 John Oxford, professor of virology at Queen Mary's Medical School, London, recently obtained permission from Sykes's descendants to exhume Sir Mark's body and perform an autopsy. If the coffin was properly sealed and the corpse has not decomposed then Oxford should be able to retrieve whole tissue from Sir Mark's lungs. To date just five samples of lung tissue containing the 1918 virus – H1N1 – have been recovered but the viral fragments were partial and incomplete. Oxford's hope is that the autopsy material from Sir Mark's lungs will enable scientists to obtain a complete sequence of the 1918 virus, thus shedding light on its origins and the mutations that made it such a deadly pathogen.
169 Lloyd George had a different theory blaming Sykes's death on nervous exhaustion and anxiety brought on by the pressure of the talks.
170 Crosby (2003), 193.
171 Ibid, 189–96.
172 *The Times*, 18 December 1918.
173 *Manchester Evening News*, 9 December 1918.
174 *Hackney Gazette*, 20–23 January and 16 April 1919.
175 *The Times*, 1 June 1919.

176 G.H. Savage, 'Special Influenza Number: part 1', *The Practitioner* (January–June 1919). C. Shaw, 'The Psychoses of Influenza', *The Practitioner* (January–June 1907).

177 For further discussion of the controversy over the origins of Economo's disease and its relationship with influenza see K. Duncan, *Hunting the 1918 Flu: One Scientist's Search for a Killer Virus* (Toronto: University of Toronto Press, 2003). Recently, influenza A antigens have been isolated from the cerebrospinal fluid of patients admitted to Japanese hospitals with encephalitis or encephalopathy, and in a Japanese survey in 2000 researchers found a marked increase in encephalitis during the influenza season, particularly among young children.

178 Report on Pandemic (1920).

179 *The Times*, 6 January 1919.

180 *The Times*, 6 January 1919.

181 Letter Harry Whellock, Cape Province, South Africa, 10 November 1918. Mullocks sale item.

182 M. Davis, *The Monster at Our Door: The Global Threat of Avian Influenza* (The New Press, 2005), 25–6.

183 A. Camus, *The Plague* (Penguin, 2002).

184 Collier (1974), 304.

185 Greenwood (1935), 289.

186 P. Fussell, *The Great War and Modern Memory* (New York: Oxford University Press, 1975).

187 BBC News online, 16 October 2005.

188 Interview with author, February 2005.

189 Interview with author, March 2005.

190 'H5N1 bird flu virus reassuringly stable: animal health chief'. Agence France Presse, 10 Jan 2008.

191 *New York Times*, 22 January 2008.

192 J.K. Taubenberger, *et al.*, 'The Next Influenza Pandemic, Can It Be Predicted?', *JAMA* (9 May 2007) 297, 18: 2025–7.

193 D. Nabarro, 'The Global State of Influenza Pandemic Preparedness', LSE Health/DFID public lecture, 10 January 2008.

194 Davis (2005).

195 J. Taubenberger, and D.M. Morens, '1918 Influenza: the Mother of All Pandemics', *Emerging Infectious Diseases* (January 2006), 12: 1, 15–22.

196 Robert G. Webster, 'H5 Influenza Viruses', in Y. Kawaoka, *Influenza Virology: Current Topics* (Caister Academic Press, 2006), 281–98.

197 P. Davies, *The Devil's Flu: The World's Deadliest Epidemic and the Scientific Hunt for the Virus that Caused It* (London: Michael Joseph, 2000).

198 The doctor had infected 16 guests on the same floor and these guests, who included a businessman and international airline crew, carried the virus to Hanoi, Singapore and Toronto.

199 A. Osterhaus, *et al.*, 'The Aetiology of SARS: Koch's postulates fulfilled', *Philosophical Transactions of the Royal Society of London* (2004) 359 B: 1081–2.

200 J. Peiris and Y. Guan, 'Confronting SARS: a view from Hong Kong', *Philosophical Transactions of the Royal Society of London* (2004) 359 B: 1075–9.

201 R. Anderson, *et al.*, 'Epidemiology, Transmission Dynamics and Control of SARS: the 2002–2003 Epidemic', *Philosophical Transactions of the Royal Society of London* (2004), 359 B 1091–105.

202 Interview with author, February 2005.

203 Interview with author, April 2008.

204 D. Morens and A.S. Fauci, 'The 1918 Pandemic: Insights for the 21st Century', *Journal of Infectious Diseases* (1 April 2007), 195: 1018–28.

205 J. Taubenberger, *et al.*, 'Initial Characterization of the 1918 "Spanish" influenza virus', *Science* (1997) 275, 1793–6.

206 'The 1918 flu virus is resurrected', *Nature* (6 October 2005), 437: 794–5.

207 Ibid.

208 J. Taubenberger and P. Palese, 'The origin and virulence of the 1918 "Spanish" influenza virus', in Y. Kawaoka, *Influenza Virology: Current Topics* (Caister Academic Press, 2006), 299–321.

209 Ibid.

210 Having said that, the fact that all of the last five pandemics were caused by just three HA subtypes – H1, H2 and H3 – suggests that there may be unappreciated biological barriers preventing H5N1 from mutating into a pandemic strain. Some scientists argue that given that the majority of the world's population have no protective immunity to H2 subtype influenza viruses, i.e. the subtype which circulated between 1957 and 1968, then that is the subtype we ought to fear. For further discussion see J. Taubenberger, *et al.*, 'The Next Influenza Pandemic, Can It Be Predicted?,' *JAMA* (9 May 2007), 297, 18: 2025–7.

211 Email to author, 25 April 2005.

212 Interview with author, March 2008.

213 Kawaoka (2006), 281–98.

214 Interview with author, March 2008.

215 H. Wang, *et al.* 'Probable limited person-to-person transmission of highly pathogenic avian influenza A (H5N1) virus in China', *Lancet* (8 April 2008), online publication. www.thelancet.com/journals.

216 Further support for passive immunization comes from a recent meta-analysis of eight trials conducted between 1918–25. In the trials patients hospitalized with Spanish influenza complicated by pneumonia were given whole blood, plasma or serum obtained from other patients recently recovered from flu and their outcomes compared with people

who had received conventional hospital care. The treated patients had a case fatality rate of 16 per cent compared to 37 per cent in the untreated patients, suggesting that anti-influenza antibodies in the blood products may have neutralized the virus or reduced viral loads below the point necessary to trigger a cytokine storm. See T.C. Luke, *et al.*, 'Meta-Analysis: Convalescent Blood Products for Spanish Influenza Pneumonia: A Future H5N1 Treatment?', *Annals of Internal Medicine* (17 October 2006) 145(8): 599–609.

217 'Bird flu pandemic "will hit UK"', BBC News online, 16 October 2005.

218 Interview with author, 29 April 2008.

219 House of Lords Science and Technology Committee – Fourth Report (December 2005). Examination of Witnesses, Questions 160–179, 27 October 2005.

220 D.K. Menon, *et al.*, 'Modelling the impact of an influenza pandemic on critical care services in England', *Anaesthesia* (2005) 60: 952–4.
J. Revill, *Everything You Need to Know About Bird Flu & What You Can Do To Prepare For It* (Rodale International, 2005), 132.

221 Interview with author, 29 April 2008.

222 Interview with author, 18 April 2005.

223 'Exercise Winter Willow – Lessons Identified', *UK Resilience*, 30 January and 19–20 February 2007.

224 House of Lords Science and Technology Committee – Fourth Report (7 December 2005). 'Pandemic Flu: a national framework for responding to an influenza pandemic', Department of Health (November 2007).

225 Another possibility is immunization with a pre-pandemic vaccine. In May 2008 the European Commission granted a licence to GlaxoSmith Kline to begin distributing Prepandrix, a pre-pandemic vaccine made from strains of H5N1 circulating in Vietnam and Indonesia. In clinical trials , GSK claimed that Prepandrix resulted in a four-fold increase in antibodies in 77–85 per cent of test subjects. So far, orders have been placed by the US, Switzerland and Finland. GSK has also donated 50m doses to the World Health Organization.

226 C. Murray, *et al.*, 'Estimation of potential global pandemic influenza mortality on the basis of vital registry data from the 1918–20 pandemic: a quantitative analysis', *Lancet* (23/30 December 2006) 368: 2211–18.

227 D. Morens, and A.S. Fauci, 'The 1918 Influenza Pandemic: Insights for the 21st Century', *Journal of Infectious Diseases* (1 April 2007) 195: 1018–28.

228 S. Cauchemez, *et al.*, 'Estimating the impact of school closure on influenza transmission from Sentinel data', *Nature* (2008) 452: 750–5.

229 'Pandemic Flu: a national framework for responding to an influenza pandemic', Department of Health (22 November, 2007), 81.

230 Interview with author, 10 April 2008.

231 *The Times*, 7 March 1919.

232 Report on Pandemic (1920), xiv.

233 Ibid, xiv.

234 For instance, Hampstead and Kensington, which were two of the wealth-iest boroughs, had the lowest epidemic death rates, but Chelsea, which was the third wealthiest borough, had a death rate almost as high as St Pancras, one of the poorest boroughs. See S. Tomkins, 'The Failure of Expertise: Public Health Policy in Britain during the 1918–19 Influenza Epidemic', *Social History of Medicine* (December 1992) 5: 435–54.

235 Johnson (2006), 100–3.

236 Hamer (1919).

237 Report on Pandemic (1920), xviii.

238 *BMJ* (5 June 1948), 1112–13.

239 M. Fletcher, *The Bright Countenance: A Personal History of Walter Morley Fletcher* (London: Hodder & Stoughton, 1957), 143.

240 Hamer (1919).

241 *BMJ*, 17 October 1925, 710.

242 Oxford Dictionary of National Biography, 40: 928–9.

selected bibliography

Primary Sources
COLLECTIONS

Caroline Playne Collection – MS1112, Senate House Library, University of London.
Richard Collier Collection – 63/5/2-7, Imperial War Museum.

ARCHIVES

Contemporary Medical Archives Centre, Wellcome Library.
'Notes on St. Marylebone Infirmary (later St. Charles Hospital) 1910–1941', compiled by Dr. Basil Hood – GC/21.
Letters from Sir John Rose Bradford and his wife Mary Bradford, largely while serving with the Royal Army Medical Corps in France, 1905–1921 – PP/BAR/Z/9 SERIES

National Archives

FD 1/529 – Influenza Research.
FD 1/533 – Medical Research Committee, Influenza – General Research in UK.
FD 1/534 – Medical Research Committee, Influenza – General research in UK.
FD 1/535 – Influenza Committee Correspondence with Local Government Board and War Office 1918–20.
FD 5/187 – Bacteriological and Pathological Observations of Influenza in France during 1918. A report by W.J. Wilson, Major RAMC, and Sergeant P. Steer RAMC.
FD 1/529 – Medical Research Committee, Influenza – Research by Colonel Cumming, Bowman and Gibson with British Forces in France.
MH 139/3 – Diaries of Sir George Newman, Chief Medical Officer 1916–20.

Secondary Sources
NEWSPAPERS/PERIODICALS

Daily Mirror
Daily News

Daily Sketch
Glasgow Herald
Hackney Gazette
Manchester Guardian
Manchester Evening News
Observer
Salford Reporter
Sheffield Independent
Times
Yorkshire Telegraph
Tatler
Spectator

OFFICIAL REPORTS

'Pandemic Flu: a national framework for responding to an influenza pandemic', Department of Health (November 2007).

Hamer, W. 'Report on Influenza' (June 1919), London County Council official publication, 210: 1963, 2.

House of Lords Science and Technology Committee – Fourth Report (7 December 2005).

Medical Research Committee. *Special Report Series*, No. 36, 'Studies of Influenza in hospitals of the British Armies in France 1918' (London: HMSO, 1919).

'Report on the Pandemic of Influenza 1918–19', Ministry of Health, Reports on Public Health and Medical Subjects, No. 4 (London: HMSO, 1920).

Niven, J. 'Annual Report of the Medical Officer of Health for 1918' (Manchester: 1919).

Parsons, H.F. and Local Government Board. 'Report on the influenza epidemic of 1889–90' (London: Eyre & Spottiswoode, 1891).

Parsons, H.F. and Local Government Board. 'Further Report and Papers on Epidemic Influenza 1889–92', (London: Eyre & Spottiswoode, 1893).

'Discussion on Influenza', *Proceedings of the Royal Society of Medicine* (November 13–14, 1918) 12: 1–102 (Longmans, Green & Co, 1919).

BOOKS/ARTICLES

Barry, J.M. *The Great Influenza: the Epic Story of the Deadliest Plague in History* (New York: Viking, 2004).

Beveridge, W.I.B. *Influenza: The Last Great Plague: an unfinished story of discovery* (London: Heinemann, 1977).

Brittain, V. *Testament of youth: an autobiographical study of the years 1900–1925* (London: Virago, 1978).

Burnet, F.M. and Clark, E. *Influenza: a survey of the last 50 years in the light of modern work on the virus of epidemic influenza* (Melbourne: Macmillan, 1942).

Byerly, C.R. *Fever of war: the influenza epidemic in the U.S. Army during World War I* (New York; London: New York University Press, 2005).

Collier, R. *The Plague of the Spanish Lady: the influenza pandemic of 1918–1919* (London: Macmillan, 1974).

Creighton, C. *A history of epidemics in Britain* (Cambridge: The University Press, 1894).

Crosby, A.W. *America's forgotten pandemic: the influenza of 1918* (Cambridge: Cambridge University Press, 2003).

Davies, P. *The Devil's Flu: The world's deadliest epidemic and the scientific hunt for the virus that caused it* (London: Michael Joseph, 2000).

Davis, M. *The Monster at Our Door: the global threat of avian influenza* (New York; London: The New Press, 2005).

DeGroot, G.J. *Blighty: British society in the era of the Great War* (London; New York: Longman, 1966).

Duncan, K. *Hunting the 1918 Flu: one scientist's search for a killer virus* (Toronto: University of Toronto Press, 2003).

Eyler, J.M. *Sir Arthur Newsholme and State Medicine, 1885–1935* (New York: Cambridge University Press, 1997).

Fletcher, M. *The Bright Countenance: a personal history of Walter Morley Fletcher* (London: Hodder & Stoughton, 1957).

Fussell, P. *The Great War and Modern Memory* (New York: Oxford University Press, 1975).

Graves, R. *Goodbye to All That: an autobiography* (London: Jonathan Cape, 1929).

Greenwood, M. *Epidemics and Crowd-diseases: an introduction to the study of epidemiology* (London: Williams & Norgate, 1935).

Grigg, J. *Lloyd George: war leader* (London: Allen Lane, 2002).

Iezzoni, L., McCullough, D.G. *Influenza 1918: the worst epidemic in American history* (New York: TV Books, 1999).

James, R. *This Time of Dying* (London: Portobello, 2006).

Johnson, N. *Britain and the 1918–19 Influenza Pandemic: a dark epilogue* (London: Routledge, 2006).

Kawaoka, Y. *Influenza Virology: current topics* (Wymondham Caister Academic Press, 2006).

Kolata, G.B. *Flu: the story of the great influenza pandemic of 1918 and the search for the virus that caused it* (London: Macmillan, 2000).

MacDonagh, M. *In London During the Great War: the diary of a journalist* (London: Eyre and Spottiswoode, 1935).

Macpherson, W.G. *Medical Services, General History* (London: H.M.S.O, 1921–24).

Marwick, A. *The Deluge: British society and the First World War* (Basingstoke: Macmillan Education, 1991).

Niven, J. *Observations on the History of Public Health Effort in Manchester* (Manchester: J. Heywood, 1923).

Owen, H. and Bell J. *Wilfred Owen, Collected Letters* (London: Oxford University Press, 1967).

Peel, C.S. *How We Lived then, 1914–1918* (London: John Lane, 1929).

Playne, C.E. *Britain Holds On, 1917, 1918* (London: Allen and Unwin, 1933).

Phillips, H. and Killingray, D. *The Spanish Influenza Pandemic of 1918–19: new perspectives* (London; New York: Routledge, 2003).

Revill, J. *Everything You Need to Know About Bird Flu & What You Can Do To Prepare For It* (London: Rodale International, 2005).

Rice, G. and Bryder, L. *Black November: the 1918 influenza pandemic in New Zealand* (Christchurch, New Zealand: Canterbury University Press, 2005).

Tanner, A. 'The Spanish Lady Comes to London', *London Journal* (2002) 2(27–8): 51–76.

Thomson, D.T. and Thomson, R. 'Influenza,' Annals of the Pickett-Thomson Research Laboratory, volumes 9 and 10 (London: Bailliere Tindall & Cox, 1934).

Tomkins, S. 'The Failure of Expertise: Public Health Policy in Britain during the 1918–19 Influenza Epidemic', *Social History of Medicine* (December 1992) 5: 435–54.

Wilson, T. *The Myriad Faces of War: Britain and the Great War, 1914–1918* (Cambridge: Polity Press, 1986).

Winter, J.M. *Sites of Memory, Sites of Mourning: the Great War in European cultural history* (Cambridge: Cambridge University Press, 1995).

Winter, J.M. *The Great War and the British People* (London: Macmillan Education, 1986).

Winter, J.M. and Robert, J.L. *Capital Cities at War, London, Paris, Berlin 1914–19* (Cambridge: Cambridge University Press, 1997).

index